Random Linear Operators

Mathematics and Its Applications *(Soviet Series)*

Managing Editor:

M. HAZEWINKEL
Centre for Mathematics and Computer Science, Amsterdam, The Netherlands

Editorial Board:

L. D. FADDEEV, *Steklov Institute of Mathematics, Leningrad, U.S.S.R.*
A. A. KIRILLOV, *Steklov Institute of Mathematics, Moscow, U.S.S.R.*
Yu. I. MANIN, *Steklov Institute of Mathematics, Moscow, U.S.S.R.*
N. N. MOISEEV, *Steklov Institute of Mathematics, Moscow, U.S.S.R.*
S. P. NOVIKOV, *Steklov Institute of Mathematics, Moscow, U.S.S.R.*
M. C. POLYVANOV, *Steklov Institute of Mathematics, Moscow, U.S.S.R.*
Yu. A. ROZANOV, *Steklov Institute of Mathematics, Moscow, U.S.S.R.*

A. V. Skorohod
Institute of Mathematics, Kiev, U.S.S.R.

Random Linear Operators

D. Reidel Publishing Company
A MEMBER OF THE KLUWER ACADEMIC PUBLISHERS GROUP
Dordrecht / Boston / Lancaster

Library of Congress Cataloging in Publication Data

CIP

Skorohod, A. V. (Anatolii Vladimirovich), 1930–
Random linear operators.

(Mathematics and its applications. Soviet series)
Translation of: Sluchainye lineinye operatory.
Bibliography: p.
Includes index.
1. Random operators. 2. Linear operators. I. Title.
II. Series: Mathematics and its applications (D. Reidel Publishing Company). Soviet series.
QA274.28.S5613 1983 519.2 83-19188
ISBN 90-277-1669-2

Published by D. Reidel Publishing Company
P.O. Box 17, 3300 AA Dordrecht, Holland

Sold and distributed in the U.S.A. and Canada
by Kluwer Academic Publishers,
190 Old Derby Street, Hingham, MA 02043, U.S.A.

In all other countries, sold and distributed
by Kluwer Academic Publishers Group,
P.O. Box 322, 3300 AH Dordrecht, Holland

Translated from the Russian by UAT Ltd.

Printed in The Netherlands

TABLE OF CONTENTS

EDITOR'S PREFACE

Approach your problems from the right end and begin with the answers. Then one day, perhaps you will find the final question.

'The Hermit Clad in Crane Feathers' in R. van Gulik's The Chinese Maze Murders.

It isn't that they can't see the solution.
It is that they can't see the problem.

G. K. Chesterton. The Scandal of Father Brown 'The Point of a Pin'.

Growing specialization and diversification have brought a host of monographs and textbooks on increasingly specialized topics. However, the "tree" of knowledge of mathematics and related fields does not grow only by putting forth new branches. It also happens, quite often in fact, that branches which were thought to be completely disparate are suddenly seen to be related.

Further, the kind and level of sophistication of mathematics applied in various sciences has changed drastically in recent years: measure theory is used (non-trivially) in regional and theoretical economics; algebraic geometry interacts with physics; the Minkowsky lemma, coding theory and the structure of water meet one another in packing and covering theory; quantum fields, crystal defects and mathematical programming profit from homotopy theory; Lie algebras are relevant to filtering; and prediction and electrical engineering can use Stein spaces. And in addition to this there are such new emerging subdisciplines as "completely integrable systems", "chaos, synergetics and large-scale order", which are almost impossible to fit into the existing classification schemes. They draw upon widely different sections of mathematics.
This program, Mathematics and Its Applications, is devoted to such (new) interrelations as exampla gratia:

- a central concept which plays an important role in several different mathematical and/or scientific specialized areas;
- new applications of the results and ideas from one area of scientific endeavor into another;
- influences which the results, problems and concepts of one field of enquiry have and have had on the development of another.

The Mathematics and Its Applications programme tries to make available a careful selection of books which fit the philosophy outlined above. With such books, which are stimulating rather than definitive, intriguing rather than encyclopaedic, we hope to contribute something towards better communication among the practitioners in diversified fields.

Because of the wealth of scholarly research being undertaken in the Soviet Union, Eastern Europe, and Japan, it was decided to devote special attention to the work emanating from these particular regions.
Thus it was decided to start three regional series under the umbrella of the main MIA programme.

The present volume is the first to appear in MIA (Soviet Series). Its subject belongs to the general field "random analysis" and it is agreeable to be able to remark that this is one of the fastest growing areas in mathematics, both as to applications and as regards the development of the theory. Some quick counting in, for example, issues of Mathematical Reviews will prove this. Also probabilistic analysis seems to have a tendency to become more and more intertwined with other parts of mathematics rather than less.

After random points, random functions (i.e. random processes), random sets (and integral geometry), it is natural to expect the subject of random operators to arise and flourish. Not only for its own sake, but as the author stresses, also because applications demand such a theory. It is the first complete book on a subject which will certainly undergo a lot of development in the next decade, and which, at the moment, presents many challenging (open) problems, besides a substantial body of applicable results.

The unreasonable effectiveness of mathematics in science ...

Eugene Wigner

Well, if you knows of a better 'ole, go to it.

Bruce Bairnsfather

What is now proved was once only imagined.

William Blake

As long as algebra and geometry proceeded along separate paths, their advance was slow and their applications limited.
But when these sciences joined company they drew from each other fresh vitality and thenceforward marched on at a rapid pace towards perfection.

Joseph Louis Lagrange

Amsterdam, July 1983

Michiel Hazewinkel

PREFACE

Random operator theory of one of the newest branches of the theory of random processes and functions: its creation is a natural step in the development of "random analysis". Among the well-known branches of this type of analysis one can include the theory of differentiation of random functions, stochastic differential equations, random sequences and series, and linear topological spaces with measure. One of the current developments is the theory of linear equations with random coefficients, which bears an obvious relationship to the general theory of linear operators. The interest in random operators has been aroused not merely by the desire to explore a specific area of research as thoroughly as possible: the need for a theory of such operators is dictated, on the one hand, by the intrinsic needs of the theory of random processes and, on the other, by problems that arise when the latter theory is applied in mechanics, theoretical physics and radio engineering. Thus, in the analysis of certain classes of controlled processes one uses conditional Markov processes whose transition probabilities are themselves random and so generate random transition operators. An extensive class of random operators is encountered when one defines different kinds of stochastic integrals in the theory of stochastic differential equations. Finally, random operators play an important role in the statistics of random processes, such as, for example, the empirical correlation operator of a process.

Linear random functions have received considerable attention in the theory of random processes. A whole series of publications are devoted to the investigation of random variables with values in an infinite-dimensional linear space [4, 5, 11, 17, 27, 28]. The first question that arises here is to characterize the distributions in a given space, i.e., to find conditions under which finite-dimensional distributions ("weak distribution") generate a measure. Suitable conditions for Hilbert and nuclear spaces are now available in the Minlos-Sazonov Theorem [15, 18]. The fact that not every weak distribution is the distribution of some random variable prompted Daletskii [12] to consider generalized random variables belonging to some extension of the original space. These generalized random variables are now being widely used. Analysis of the random operators actually utilized in probability theory has shown that they may be regarded only as generalized variables in the space of linear operators. We are referring here to operators generated by stochastic integrals, which figure significantly in the theory of stochastic differential equations.

This monograph presents, for the first time, a complete account of the theory of these generalized operators, how they may be defined, conditions under which they can be viewed as generalized random variables in operator space, as well as multiplication and limit procedures in the space of random operators,(not counting various publications of the author in the journal literature [24, 25]).

Besides the general questions occasioned by the fact that random operators are random elements of an infinite--dimensional space, random operator theory can be applied to specific problems stemming from the fact that these random elements are operators. One type of problem is the investigation of the solutions of random operator equations and the spectra of random operators. Closely connected with this is the investigation of the spectra of random matrices of order approaching infinity. This problem was first formulated by Marchenko and Pastur, in the context of actual physical models [14, 16]. Several general theorems concerning the limiting behavior of the spectrum have been proved by Girko [10], who has also established a highly interesting description of the joint distribution of the eigenvectors and eigenvalues of a symmetric random matrix.

In Chapter 2, the spectral properties of certain classes of random operators are studied, as are solutions of operator equations. Some of the results of this chapter were published in [25].

The source of the second type of problem is the fact that the set of linear operators is a noncommutative ring. One can therefore consider products of independent operators and the continuous analog of these products - multiplicative processes.

There exists a considerable literature on products of independent elements of groups and their limiting behavior (see the surveys [11, 19]). However, there is little information that one can deduce from this literature, even for matrices, since the group of invertible matrices is non-compact. As for the group of nonsingular operators - it is not even locally compact. For this reason, no limit theorems have as yet been obtained and we are not even sure how to attack such problems as the asymptotic behavior of the distribution of a product. The situation is simple with regard to the convergence of infinite products of matrices and operators [26]. The most general results are presented in Chapter 3, § 8, which also considers operator-valued martingales of discrete and continuous arguments. Discrete martingales are a convenient tool in studying products of independent random operators, while continuous martingales are needed in the construction of stochastic integrals (Chapter 4), to be used in the investigation of multiplicative processes as solutions of linear stochastic operator equations. The matrix version of multiplicative processes was introduced by the present author in [21] and later studied by Butsan in a series of papers, summarized in [3]. Simultaneously, random multiplicative semigroups of probability operators made their appearance in the theory of random processes, in the context of research on conditional Markov processes. A description of these semigroups for the discontinuous component was given by the author in [22], where they were presented as solutions of a linear stochastic equation. General multiplicative operator functionals as solutions of linear equations were constructed for various classes of Markov processes by Daletskii and his students [2].

The present author introduced the concept of stochastic semigroup with independent values [24, 29] developing a linear stochastic differential equation for continuous semigroups. A large class of stochastic semigroups (without the continuity assumption) was studied by Butsan [3].
The most general results on linear stochastic operator equations and continuous stochastic semigroups are described in Chapter 5.

LIST OF NOTATIONS

$\mathbf{X}$ - separable Hilbert space
$\mathfrak{B}$ - σ-algebra of Borel subsets of $\mathbf{X}$ (elements of $\mathbf{X}$ are denoted by lower-case Latin letters)
(x, y) - scalar product of elements $x, y \in \mathbf{X}$
$|x|$ - norm of $x \in \mathbf{X}$
$\{e_k, k = 1, 2, \dots\}$ - orthonormal basis in $\mathbf{X}$, i.e., $(e_k, e_j) = \delta_{kj} = \begin{cases} 1, & k = j; \\ 0, & k \neq j \end{cases}$

$\mathbf{R}$ - the real line (its elements are denoted by lower-case Greek letters)
$\{\Omega, \mathfrak{S}, \mathbf{P}\}$ - probability space; events (elements of $\mathfrak{S}$) are denoted by Greek capitals
$\xi(\omega)$ - random variable (r.v.)
$\mathbf{M}\xi(\omega)$ - expectation of the above r.v.
$\mathbf{D}\xi(\omega)$ - variance
$\mathbf{L}(\mathbf{X})$ - set of (bounded) linear operators from $\mathbf{X}$ to $\mathbf{X}$
For $A \in \mathbf{L}(\mathbf{X})$ $\|A\| = \sup\limits_{|x| \leqslant 1} |Ax|$
For $A \in \mathbf{L}(\mathbf{X})$ the operator A^* for which $\forall_{x,y} \in \mathbf{X}$, $(Ax, y) = (x, A^*y)$, is called the adjoint of A
If $A = A^*$, then A is symmetric
If $A = A^*$ and $(Ax, x) \geqslant 0$ $\forall x \in \mathbf{X}$, then $A \in \mathbf{L}_+(\mathbf{X})$
An operator A is nuclear if $A \in \mathbf{L}_+(\mathbf{X})$ and $\operatorname{sp} A = \Sigma (Ae_k, e_k) < \infty$ relative to some basis $\{e_k\}$
A is a Hilbert-Schmidt operator if A^*A is nuclear
$\mathbf{X}(\Omega)$ - set of functions $x(\omega)$ from Ω to $\mathbf{X}$ ($\mathbf{X}$-valued r.v.'s) which are measurable relative to the pair $(\mathfrak{B}, \mathfrak{S})$
$\mathbf{L}(\Omega, \mathbf{X})$ - set of mappings $A(\omega)$ from Ω to $\mathbf{L}(\mathbf{X})$, such that $(A(\omega)x, y)$ is $\mathfrak{S}$-measurable for all $x, y \in \mathbf{X}$
$\alpha \wedge \beta$ - smaller of the numbers α and β
$\alpha \vee \beta$ - larger of the numbers α and β

CHAPTER 1

RANDOM OPERATORS IN HILBERT SPACE

1. Basic Definitions

1.1. Strong Random Operator

The simplest example of a random operator is the stochastic integration operator. Consider the Hilbert space $L_2[0, 1]$, a Wiener process $w(t)$ on $[0, 1]$ given $\varphi \in L_2[0, 1]$, we put

$$A(\omega)\varphi(t) = \int_0^t \varphi(s)\, dw(s),$$

where $A(\omega)\varphi(t)$ is a continuous random function and so belongs to $\mathbf{L}_2[0, 1]$ with probability 1. Hence the operator $A(\omega)$ maps functions in $\mathbf{L}_2[0, 1]$ to random functions in $\mathbf{L}_2[0, 1]$. However, $A(\omega)$ does not belong to $\mathbf{L}(\Omega, X)$. Indeed, let

$$\varphi_n(s) = \sqrt{n}\left[w\left(\frac{k}{n}\right) - w\left(\frac{k-1}{n}\right)\right], \quad \frac{k-1}{n} \leqslant s < \frac{k}{n}, \quad k = 1, \ldots, n.$$

Then

$$\int_0^1 \varphi_n^2(s)\, ds = \sum_{k=1}^{n}\left[w\left(\frac{k}{n}\right) - w\left(\frac{k-1}{n}\right)\right]^2.$$

This expression tends to 1 with probability 1 as $n \to \infty$. At the same time, using the definition of the integral of a step function and extending it in the natural way to functions depending on ω, we obtain

$$\int_0^1 |A(\omega)\varphi(s)|^2\, ds = \sum_{k=1}^{n} \int_{\frac{k-1}{n}}^{\frac{k}{n}} \left(\sqrt{n}\sum_{j=1}^{k-1}\left[w\left(\frac{j}{n}\right) - w\left(\frac{j-1}{n}\right)\right]^2 +\right.$$

$$+\sqrt{n}\left(w\left(\frac{k}{n}\right)-w\left(\frac{k-1}{n}\right)\right)\left(w(t)-w\left(\frac{k-1}{n}\right)\right)\Big)^2 dt \geqslant$$

$$\geqslant \sum_{k=1}^{n}\int_{\frac{k-1}{n}}^{\frac{k}{n}}\left\{\frac{1}{2}n\left(\sum_{j=1}^{k-1}\left(w\left(\frac{j}{n}\right)-w\left(\frac{j-1}{n}\right)\right)^2\right)^2-\right.$$

$$\left.-2n\left(w\left(\frac{k}{n}\right)-w\left(\frac{k-1}{n}\right)\right)^2\left(w(t)-\omega\left(\frac{k-1}{n}\right)\right)^2\right\}dt$$

(where we have made use of the inequality $(\alpha+\beta)^2 \geqslant \frac{1}{2}\alpha^2 - 2\beta^2$).

Since $\mathbf{M}\left(w\left(\frac{k}{n}\right)-w\left(\frac{k-1}{n}\right)\right)^2\left(w(t)-w\left(\frac{k-1}{n}\right)\right)^2 = O\left(\frac{1}{n^2}\right)$,

the expression

$$\sum_{k=1}^{n}\int_{\frac{k-1}{n}}^{\frac{k}{n}}\left(w\left(\frac{k}{n}\right)-w\left(\frac{k-1}{n}\right)\right)^2\left(w(t)-w\left(\frac{k-1}{n}\right)\right)^2 dt$$

is bounded in probability. Further,

$$\sum_{k=1}^{n}\int_{\frac{k-1}{n}}^{\frac{k}{n}} n\left(\sum_{j=1}^{k-1}\left(w\left(\frac{j}{n}\right)-w\left(\frac{j-1}{n}\right)\right)^2\right)^2 dt =$$

$$=\sum_{k=1}^{n}\left(\sum_{j=1}^{k-1}\left(w\left(\frac{j}{n}\right)-w\left(\frac{j-1}{n}\right)\right)^2\right)^2 \sim \sum_{k=1}^{n}\left(\frac{k-1}{n}\right)^2 \sim \frac{n}{3}\to\infty.$$

We have thus constructed a sequence of functions $\dot{\varphi}_n(t)$ which are bounded in norm and such that $A(\omega)\varphi(t)$ converges in norm to ∞. This would be impossible if $A(\omega)$ were in $\mathbf{L}(\Omega, \mathbf{X})$. Our example shows that the set of random operators $\mathbf{L}(\Omega, \mathbf{X})$ is inadequate. If $A(\omega)\in\mathbf{L}(\Omega, \mathbf{X})$, then $\forall x\in\mathbf{X}$ $A(\omega)x\in\mathbf{X}(\Omega)$. Moreover, $A(\omega)x$ satisfies the following conditions:

(S1) $\forall x_1, x_2\in\mathbf{X}$, α, $\beta\in\mathbf{R}$

$$\mathbf{P}\{A(\omega)(\alpha x_1+\beta x_2)=\alpha A(\omega)x_1+\beta A(\omega)x_2\}=1;$$

(S2) $A(\omega)x$ is continuous in x in probability:

$$\forall\varepsilon>0\ \lim_{x_n\to x}\mathbf{P}\{|A(\omega)x_n-A(\omega)x|>\varepsilon\}=0.$$

Let $A(\omega)x$ be a mapping of $\mathbf{X}$ into $\mathbf{X}(\Omega)$ satisfying conditions (S1) and (S2). Then $A(\omega)x$ defines a strong random operator. The set of strong random operators is denoted by $\mathbf{L}_s(\Omega, \mathbf{X})$. Operators $A_1(\omega)$ and $A_2(\omega)$ in this set are considered identical if

$$\forall x \in \mathbf{X} \quad \mathbf{P}\{A_1(\omega)x = A_2(\omega)x\} = 1. \tag{1.1}$$

Subject to this convention, any operator $A(\omega) \in \mathbf{L}(\Omega, \mathbf{X})$ may be identified with a certain strong operator, so that

$$\mathbf{L}_s(\Omega, \mathbf{X}) \supset \mathbf{L}(\Omega, \mathbf{X}).$$

It is clear from the example considered above that this inclusion is proper, since it is readily seen that the operator constructed above is strong.

1.2. Weak Random Operator

Let $A(\omega) \in \mathbf{L}_s(\Omega, \mathbf{X})$. Then the expression $(A(\omega)x, y)$ defines a mapping of $\mathbf{X} \times \mathbf{X}$ into $\mathbf{R}(\Omega)$ which satisfies the following conditions:

(W1) $\forall x_1, x_2, y_1, y_2 \in \mathbf{X},\ \alpha_1, \alpha_2, \beta_1, \beta_2 \in \mathbf{R}$

$$\mathbf{P}\left\{(A(\omega)(\alpha_1 x_1 + \alpha_2 x_2), \beta_1 y_1 + \beta_2 y_2) = \sum_{i,j=1}^{2} \alpha_i \beta_j (A(\omega)x_i, y_j)\right\} = 1;$$

(W2) $(A(\omega)x, y)$ is continuous in probability on $\mathbf{X} \times \mathbf{X}$: $\forall \varepsilon > 0$

$$\lim_{x_n \to x, y_n \to y} \mathbf{P}\{|(A(\omega)x_n, y_n) - (A(\omega)x, y)| > \varepsilon\} = 0.$$

Given a mapping $(A(\omega)x, y)$ of $\mathbf{X} \times \mathbf{X}$ into $\mathbf{R}(\Omega)$, which satisfies conditions (W1) and (W2), we say that this defines a weak random operator $A(\omega)$. The set of all weak random operators is denoted by $\mathbf{L}_w(\Omega, \mathbf{X})$. Operators $A_1(\omega)$ and $A_2(\omega)$ are identical if

$$\forall x, y \in \mathbf{X}\ \mathbf{P}\{(A_1(\omega)x, y) = (A_2(\omega)x, y)\} = 1. \tag{1.2}$$

It is easy to see that

$$\mathbf{L}_w(\Omega, \mathbf{X}) \supset \mathbf{L}_s(\Omega, \mathbf{X}) \supset \mathbf{L}(\Omega, \mathbf{X}).$$

Note that the identification of operators $\mathbf{L}_w(\Omega, \mathbf{X})$ and $\mathbf{L}_s(\Omega, \mathbf{X})$ is compatible: if $A_1(\omega), A_2(\omega) \in \mathbf{L}_s(\Omega, \mathbf{X})$ and (1.2) is satisfied, then for any countable subset $\mathbf{X}_1 \subset \mathbf{X}$ we obtain

$$\mathbf{P}\{(A_1(\omega)x - A_2(\omega)x, y) = 0, \ y \in \mathbf{X}_1\} = 1.$$

Since $A_1(\omega)x - A_2(\omega)x \in \mathbf{X}$, this implies (1.1). On the other hand, if $A_1(\omega), A_2(\omega) \in \mathbf{L}_w(\Omega, \mathbf{X})$, then it follows from (1.2) that

$$\mathbf{P}\{A_1(\omega) = A_2(\omega)\} = 1.$$

Thus, a bounded random operator is defined for all $\omega \in \Omega$; for a strong random operator, only the result of applying the operator to an arbitrary (nonrandom) element $x \in \mathbf{X}$ is defined; and finally, given a weak operator, we can define only the scalar product $(A(\omega)x, y)$ for any $x, y \in \mathbf{X}$. Since weak operators are the most simply defined, it is natural that any operator will first be defined in weak form, and it will then be checked for membership in the smaller set $\mathbf{L}_s(\Omega, \mathbf{X})$ or $\mathbf{L}(\Omega, \mathbf{X})$. Below we shall present two conditions for an operator to be strong or bounded.

Theorem 1. Let $\{e_k\}$ be a basis in $\mathbf{X}$, $A(\omega) \in \mathbf{L}_w(\Omega, \mathbf{X})$. Then $A(\omega) \in \mathbf{L}_s(\Omega, \mathbf{X})$ if an only if

$$\forall x \in \mathbf{X} \quad \mathbf{P}\left\{\sum_{k=1}^{\infty} (A(\omega)x, e_k)^2 < \infty\right\} = 1.$$

Proof. Necessity. If $A(\omega)x \in \mathbf{X}$, then for all $\omega \in \Omega$

$$A(\omega)x = \sum_{k=1}^{\infty} (A(\omega)x, e_k)e_k;$$

$$|A(\omega)x|^2 = \sum_{k=1}^{\infty} (A(\omega)x, e_k)^2 < \infty.$$

Sufficiency. Let us put

$$\hat{A}(\omega)x = \sum_{k=1}^{\infty} (A(\omega)x, e_k)e_k.$$

Then $\hat{A}(\omega)x$ satisfies condition (S1) by virtue of the fact that $(A(\omega)x, y)$ satisfies condition (W1). In addition, with probability 1,

$$(\hat{A}(\omega)x,\ y) = \sum_{k=1}^{\infty} (A(\omega)x,\ e_k)(e_k,\ y) =$$

$$= \lim_{n\to\infty}\left(A(\omega)x,\ \sum_{k=1}^{n}(e_k,\ y)e_k\right) = (A(\omega)x,\ y)$$

in view of condition (W2) and the fact that $y = \sum_{k=1}^{\infty}(e_k, y)e_k$. Hence

$$\mathbf{P}\{(\hat{A}(\omega)x,\ y) = (A(\omega)x,\ y)\} = 1.$$

We now verify that $\hat{A}(\omega)x$ satsifies condition (S2). To this end it suffices to show that

$$\lim_{\alpha\to\infty}\sup_{|x|\leqslant 1}\mathbf{P}\{|\hat{A}(\omega)x| > \alpha\} = 0. \tag{1.3}$$

Indeed, from (1.3) we obtain

$$\mathbf{P}\{|\hat{A}(\omega)x_n - \hat{A}(\omega)x| > \varepsilon\} \leqslant \sup_{|y|\leqslant 1}\mathbf{P}\left\{|\hat{A}(\omega)y| > \frac{\varepsilon}{|x_n - x|}\right\} \to 0$$

as $|x_n - x| \to 0$. We now note that (1.3) is equivalent to the following statement: for any $\varepsilon > 0$, there exists a sphere $S_\delta(x_0)$ in $\mathbf{X}$ (centered at x_0, of radius δ) and a number $\alpha > 0$ such that

$$\sup_{x\in S_\delta(x_0)}\mathbf{P}\{|\hat{A}(\omega)x| > \alpha\} \leqslant \varepsilon. \tag{1.4}$$

This follows from (1.3), since

$$\sup_{x\in S_\delta(x_0)}\mathbf{P}\{|\hat{A}(\omega)x| > \alpha\} \leqslant \sup_{x\in S_1(0)}\mathbf{P}\left\{|\hat{A}(\omega)x| > \frac{\alpha}{|x_0| + \delta}\right\}.$$

If (1.4) is valid, then we obtain for $x \in S_1(0)$

$$\mathbf{P}\{|\hat{A}(\omega)x| > \lambda\} = \mathbf{P}\{|\hat{A}(\omega)\delta x| > \lambda\delta\} =$$

$$= \mathbf{P}\{|\hat{A}(\omega)\delta x| - |\hat{A}(\omega)x_0| > \lambda\delta - |\hat{A}(\omega)x_0|\} \leqslant$$

$$\leqslant \mathbf{P}\{|\hat{A}(\omega)(x_0 + \delta x)| > \lambda\delta - |\hat{A}(\omega)x_0|\} \leqslant$$

$$\leqslant \mathbf{P}\left\{\left|\hat{A}(\omega)x_0\right| > \frac{\lambda\delta}{2}\right\} + \sup_{y\in S_\delta(x_0)}\mathbf{P}\left\{\left|\hat{A}(\omega)y\right| > \frac{\lambda\delta}{2}\right\}.$$

Now, letting $\lambda \to \infty$, we get

$$\varlimsup_{\lambda\to\infty} \sup_{x\in S_1(0)} \mathbf{P}\{|\hat{A}(\omega)x| > \lambda\} \leqslant \varepsilon,$$

for any $\varepsilon > 0$.

Now assume that (1.4) is not satisfied. This means that there exists $\varepsilon > 0$, such that, for any sphere $S_\delta(x_0)$ and $\alpha > 0$,

$$\sup_{x\in S_\delta(x_0)} \mathbf{P}\{|\hat{A}(\omega)x| > \alpha\} > \varepsilon. \tag{1.5}$$

Let $\mathbf{P}\{|\hat{A}(\omega)x'| > \alpha\} < \varepsilon$. Then, for all sufficiently large n,

$$\mathbf{P}\left\{\sum_{k=1}^{n} (\hat{A}(\omega)x', e_k)^2 > \alpha^2\right\} > \varepsilon.$$

Thus, thanks to the stochastic continuity of $\sum_{k=1}^{n} (\hat{A}(\omega)\times x, e_k)^2$, we can find a number δ_1 such that, whenever $|x' - x| < \delta_1$, we again have the inequality,

$$\mathbf{P}\left\{\sum_{k=1}^{n} (\hat{A}(\omega)x, e_k)^2 > \alpha^2\right\} > \varepsilon.$$

But then, for $x \in S_{\delta_1}(x')$,

$$\mathbf{P}\{|\hat{A}(\omega)x|^2 > \alpha\} > \varepsilon.$$

Thus, if (1.5) is valid, there exists a sphere $S_\delta(x_1) \subset S_\delta(x_0)$, such that for all $x \in S_{\delta_1}(x_1)$

$$\mathbf{P}\{|\hat{A}(\omega)x| > \alpha\} > \varepsilon.$$

Using this fact, we can construct a sequence of nested spheres $S_{\delta_k}(x_k)$ in such a way that $\delta_k \downarrow 0$ and $x \in S_{\delta_k}(x_k)$

$$\mathbf{P}\{|\hat{A}(\omega)x| > k\} > \varepsilon.$$

Then, for $\bar{x} \in \bigcap_k S_{\delta_k}(x_k)$ we obtain

$$\mathbf{P}\{|\hat{A}(\omega)\bar{x}| = +\infty\} > \varepsilon,$$

and this is impossible.

Remark. In our proof of (1.3) we have used an idea of Banach to prove that a weakly bounded sequence is bounded in norm. As we shall frequently have occasion to utilize this type of argument, we now state it as a separate lemma; the proof is essentially a repetition of the proof of (1.3).

Lemma 1. Suppose we are given a sequence of random functions $\psi_n(x)$ on $\mathbf{X}$, satisfying the following conditions:

(1) $\psi_n(x) \geqslant 0$ and $\forall x, y \in \mathbf{X}\ \psi_n(x+y) \leqslant \psi_n(x) + \psi_n(y)$ with probability 1;

(2) $\forall n \lim_{\alpha\to\infty} \sup_{|x|\leqslant 1} \mathbf{P}\{\psi_n(x) > \alpha\} = 0$;

(3) $\forall x \in \mathbf{X}\quad \psi_n(x) \uparrow \psi(x)$ with probability 1.

Then

$$\lim_{\alpha\to\infty} \sup_{|x|\leqslant 1} \mathbf{P}\{\psi(x) > \alpha\} = 0.$$

The role of $\psi(x)$ in the proof of the theorem is played by $|A(\omega)x|$,

$$\psi_n(x) = \sqrt{\sum_{k=1}^{n} (\hat{A}(\omega)x, e_k)^2}.$$

Theorem 2. Let $A(\omega) \in \mathbf{L}_s(\Omega, \mathbf{X})$ and let $\mathbf{X}_1$ be a countable set, dense in $S_1(0)$, such that $\alpha_1 x_1 + \alpha_2 x_2 \in \mathbf{X}_1$ if $x_1, x_2 \in \mathbf{X}_1$, α_1, α_2 are rational numbers and $|\alpha_1 x_1 + \alpha_2 x_2| \leqslant 1$. Suppose that

$$\mathbf{P}\{\sup_{x\in X_1} |A(\omega)x| < \infty\} = 1.$$

Then $A(\omega) \in \mathbf{L}(\Omega, \mathbf{X})$.

Proof. Let $\Lambda \in \mathfrak{S}$ be the set of all ω for which $\sup_{x\in\tilde{\mathbf{X}}_1} |A(\omega)x| < \infty$, $A(\omega)(\alpha x + \beta y) = \alpha A(\omega)x + \beta A(\omega)y$ for all rational α, β and $x, y \in \mathbf{X}_1$. Since $\mathbf{X}_1$ is countable, $\mathbf{P}(\Lambda) = 1$. Let $\tilde{\mathbf{X}}$ denote the linear hull of $\mathbf{X}_1$ over the field of rationals. If $\xi(\omega) = \sup_{x\in\mathbf{X}_1} |A(\omega)x|$, then for $x, y \in \tilde{\mathbf{X}}$

$$|A(w)x - A(\omega)y| \leqslant \xi(\omega)|x-y|.$$

For any $x \in \mathbf{X}$ and $\omega \in \Lambda$, we put

$$\hat{A}(\omega)x = \lim_{n\to\infty} A(\omega)x_n,$$

where $x_n \in \tilde{\mathbf{X}}$, $x_n \to x$. The existence of this limit follows from the inequality $|A(\omega) x_n - A(\omega) x_m| \leqslant \xi(\omega) |x_n - x_m|$. It is also obvious that the value of $\hat{A}(\omega) x$ is independent of the choice of the sequence $x_n \to x$. It is readily checked that if $\omega \in \Lambda$ then $\hat{A}(\omega)(\alpha x + \beta y) = \alpha \hat{A}(\omega) x + \beta \hat{A}(\omega) y$ and $|\hat{A}(\omega) x| \leqslant \hat{\xi}(\omega) |x|$. Thus, $\hat{A}(\omega) \in \mathbf{L}(\Omega, \mathbf{X})$. Further, if $x \in \tilde{\mathbf{X}}$ then $A(\omega) x = \hat{A}(\omega) x$ for $\omega \in \Lambda$, i.e.,

$$\mathbf{P}\{A(\omega) x = \hat{A}(\omega) x\} = 1.$$

Using the fact that $A(\omega) x$ and $\hat{A}(\omega) x$ are continuous in probability as functions of x, we confirm that the last relationship is valid for any x. Thus $A(\omega) = \hat{A}(\omega)$. ■

Remark 1. Let $\mathbf{X}_1$ be as in Theorem 2 and let $A(\omega) \in \in \mathbf{L}_w(\Omega, \mathbf{X})$.

If $\mathbf{P}\{\sup_{x \in \mathbf{X}_1, y \in \mathbf{X}_1} (A(\omega) x, y) < \infty\} = 1,$

then $A(\omega) \in \mathbf{L}(\Omega, \mathbf{X})$.

The proof is analogous to that of Theorem 2.

Remark 2. Let $A(\omega) \in \mathbf{L}_w(\Omega, \mathbf{X})$ and suppose that, relative to some basis $\{e_k\}$ the following condition holds:

$$\mathbf{P}\left\{\sum_{i,k} (A(\omega) e_i, e_k)^2 < \infty\right\} = 1.$$

Then $A(\omega) \in \mathbf{L}(\Omega, \mathbf{X})$.

Indeed, since

$$|A(\omega) x|^2 = \sum_k (A(\omega) x, e_k)^2 = \sum_k \left[\sum_i (A(\omega) e_i, e_k)(x, e_i)\right]^2 \leqslant$$

$$\leqslant \sum_k \left\{\sum_i (A(\omega) e_i, e_k)^2 \sum_i (x, e_i)^2\right\} = \sum_{k,i} (A(\omega) e_i, e_k)^2 |x|^2,$$

the assumptions of the theorem are valid. The operator $A(\omega)$ is not only bounded but even a Hilbert-Schmidt operator for almost all ω, since

$$\operatorname{sp} A(\omega) A^*(\omega) = \sum_{i,k} (A(\omega) e_i, e_k)^2 < \infty$$

with probability 1.

1.3. Product of Random Operators

If $A(\omega)$ and $B(\omega)$ are in $\mathbf{L}(\Omega, \mathbf{X})$, then also $A(\omega) \times \times B(\omega) \in \mathbf{L}(\Omega, \mathbf{X})$, and then, for any basis $\{e_k\}$ we have

$$(A(\omega) B(\omega) x, y) = \sum_{k=1}^{\infty} (B(\omega) x, e_k)(A(\omega) e_k, y). \qquad (1.6)$$

It is natural that this relationship is used in order to define the product of weak operators. We first establish an auxiliary proposition.

Lemma 2. If $A(\omega)$ and $B(\omega)$ are in $\mathbf{L}_w(\Omega, \mathbf{X})$ and for some basis $\{e_k\}$ the series

$$\sum_{k=1}^{\infty} (B(\omega) x, e_k)(A(\omega) e_k, y) = \psi(x, y), \quad \forall x, y \in \mathbf{X}$$

is convergent in probability, then the function $\psi(x, y)$ is continuous in probability with respect to x and y.

Proof. Since $\psi(x, y)$ is a bilinear function, it will suffice to prove that

$$\lim_{\alpha \to \infty} \sup_{|x| \leqslant 1, |y| \leqslant 1} \mathbf{P}\{|\psi(x, y)| > \alpha\} = 0. \qquad (1.7)$$

Using the equality

$$\psi(x, y) = \psi(x_0, y_0) + \delta\psi\left(x_0, \frac{y - y_0}{\delta}\right) + $$
$$+ \delta\psi\left(\frac{x - x_0}{\delta}, y_0\right) + \delta^2\psi\left(\frac{x - x_0}{\delta}, \frac{y - y_0}{\delta}\right),$$

we see that, in turn, to prove (1.7) it will suffice to show that

$$\lim_{\alpha \to \infty} \sup_{|x| \leqslant 1} \mathbf{P}\{|\psi(x, y)| > \alpha\} = 0;$$
$$\lim_{\alpha \to \infty} \sup_{|y| \leqslant 1} \mathbf{P}\{|\psi(x, y)| > \alpha\} = 0, \qquad (1.8)$$

and to establish the existence of spheres $S_\delta(x_0)$ and $S_\delta(y_0)$ such that

$$\lim_{\alpha \to \infty} \sup_{x \in S_\delta(x_0)} \sup_{y \in S_\delta(y_0)} \mathbf{P}\{|\psi(x, y)| > \alpha\} = 0. \qquad (1.9)$$

Both relationships in (1.8) follow from Lemma 1 and also from the inequality

$$|\psi(x,\ y)| \leqslant \lim_{n\to\infty} \psi_n(x,\ y) < \infty;$$

where

$$\psi_n(x,\ y) = \sup_{m\leqslant n} \left| \sum_{k=1}^{m} (B(\omega)\,x,\ e_k)(A(\omega)\,e_k,\ y) \right|.$$

If (1.9) is false and there exists $\varepsilon > 0$ such that

$$\inf_{\alpha>0}\ \inf_{\delta>0}\ \inf_{x_0, y_0}\ \sup_{x\in S_\delta(x_0)}\ \sup_{y\in S_\delta(y_0)} \mathbf{P}\{|\psi(x,\ y)| > \alpha\} < \varepsilon, \qquad (1.10)$$

then, just as in the proof of Theorem 1, we can construct a sequence $n_k \to \infty$ and nested spheres $S_{\delta_k}(x_k) \subset S_{\delta_{k-1}}(x_{k-1})$, $S_{\delta_k}(y_k) \subset S_{\delta_{k-1}}(y_{k-1})$ $(\delta_k \downarrow 0)$, such that $\forall\, x \in S_{\delta_k}(x_k),\ y \in S_{\delta_k}(y_k)$

$$\mathbf{P}\{\psi_{n_k}(x,\ y) > k\} > \varepsilon.$$

Then, if $\bar{x} \in \bigcap S_{\delta_k}(x_k),\ \bar{y} \in \bigcap S_{\delta_k}(y_k)$ we obtain

$$\forall\, k \quad \mathbf{P}\{\psi_{n_k}(\bar{x},\ \bar{y}) > k\} > \varepsilon,$$

which is impossible.

Definition. Let $A(\omega)$ and $B(\omega)$ be in $\mathbf{L}_w(\Omega, \mathbf{X})$, and suppose that for some basis $\{e_k\}$ the series

$$\sum_{k=1}^{\infty} (B(\omega)\,x,\ e_k)(A(\omega)\,e_k,\ y)$$

converges in probability for all $x, y \in X$. Then the sum of the series defines a weak operator $A(\omega)B(\omega)$ in accordance with (1.6).

Let $A(\omega) \in \mathbf{L}_w(\Omega, \mathbf{X})$. The operator $A^*(\omega) \in \mathbf{L}_w(\Omega,\ \mathbf{X})$ such that

$$\forall\, x,\ y \in \mathbf{X} \quad (A(\omega)\,x,\ y) = (A^*(\omega)\,y,\ x),$$

is called the operator adjoint to $A(\omega)$. Its existence follows from the fact that the function $(A(\omega)\,y,\ x)$ satisfies conditions (W1) and (W2).

Lemma 3. Let $A^*(\omega)$ and $B(\omega)$ be members of $\mathbf{L}_s(\Omega, \mathbf{X})$. Then $A(\omega)\,B(\omega)$ is defined.

Indeed, the convergence of the series (1.6) follows from the inequality

$$|(B(\omega)x, e_k)(A(\omega)e_k, y)| \leqslant \frac{1}{2}[(B(\omega)x, e_k)^2 + (A^*(\omega)y, e_k)^2]$$

and from the convergence of the series

$$\sum_{k=1}^{\infty}(B(\omega)x, e_k)^2, \sum_{k=1}^{\infty}(A^*(\omega)y, e_k)^2$$

(see Theorem 1).

Let $A(\omega) \in \mathbf{L}_w(\Omega, \mathbf{X})$. Then there exists a (nonrandom) operator C such that $CA(\omega)C \in \mathbf{L}(\Omega, \mathbf{X})$. Let us assume that C is a symmetric operator such that $Ce_k = \sigma_k e_k$, then $(CA(\omega)Ce_i, e_k) = (A(\omega)Ce_i, Ce_k) = \sigma_i\sigma_k(A(\omega)e_i, e_k)$.
Now choose σ_i in such a way that the series

$$\sum_{i,k}(A(\omega)e_i, e_k)\sigma_i^2\sigma_k^2$$

is convergent with probability 1. To do this it is enough to ensure that

$$\sum_{i,k}\mathbf{P}\left\{(A(\omega)e_i, e_k)^2 > \frac{1}{k^2 i^2 \sigma_i^2 \sigma_k^2}\right\} < \infty.$$

Then the fact that $CA(\omega)C$ is bounded follows from Remark 2.

2. Characteristic Functions of Random Operators

2.1. Definition

Let $\langle u \circ v\rangle$, $u, v \in \mathbf{X}$ denote the operator in $\mathbf{L}(\mathbf{X})$, such that

$$\langle u \circ v\rangle x = (u, x)v.$$

It is clear that for any operator $A \in \mathbf{L}(\mathbf{X})$

$$A\langle u \circ v\rangle = \langle u \circ Av\rangle.$$

This formula can be extended to random operators as well. Indeed, if $A(\omega) \in \mathbf{L}_w(\Omega, \mathbf{X})$, then $(A(\omega)\langle u \circ v\rangle x, y) =$
$= (\langle u \circ A(\omega)v\rangle x, y) = (u, x)(A(\omega)v, y)$.
is defined for all $x, y \in \mathbf{X}$. The trace of the operator $\langle u \circ v\rangle$ is defined by

$$\mathrm{sp}\,\langle u \circ v\rangle = \Sigma\,(\langle u \circ v\rangle e_k,\ e_k) = \Sigma\,(u,\ e_k)(v,\ e_k) = (u,\ v).$$

Thus, we can also speak of

$$\mathrm{sp}\,A(\omega)\,\langle u\ \ v\rangle = (A(\omega)\,v,\ u).$$

Let $\mathbf{L}_0$ be the set of operators C of the form

$$C = \sum_{k=1}^{n} \langle u_k \circ v_k\rangle,\ u_k,\ v_k \in \mathbf{X},\ k = 1,\ \ldots,\ n,\ n = 1,\ \ldots .$$

Such operators will be termed degenerate. For any $C \in \mathbf{L}_0$, $A(\omega) \in \mathbf{L}_w(\Omega, \mathbf{X})$, the value of sp. $A(\omega)\,C$ is defined. The function

$$\chi_A(C) = \mathbf{M} \exp\{i\,\mathrm{sp}\,A(\omega)\,C\} \qquad (1.11)$$

is called the characteristic function of the random operator $A(\omega)$. It satisfies the following conditions:

(C1) $\chi_A(C)$ is a positive-definite as a function of C: for any $C_1, \ldots, C_n \in \mathbf{L}_0$ and complex numbers $\theta_1, \ldots, \theta_n$,

$$\sum_{k,j=1}^{n} \chi_A(C_k - C_j)\,\theta_k\bar{\theta}_j = \mathbf{M}\left|\sum_{k=1}^{n} \theta_k \exp\{i\,\mathrm{sp}\,A(\omega)\,C_k\}\right|^2 \geqslant 0;$$

(C2) $\chi_A(\langle x \circ y\rangle)$ is jointly continuous on $\mathbf{X} \times \mathbf{X}$

Theorem 1. Let $\chi_A(C)$ be a complex-valued function defined on $\mathbf{L}_0$, for which conditions (C1) and (C2) are satisfied, i.e., it is positive-definite and $\chi_A(\langle x \circ y\rangle)$ is continuous in x, y at the point zero. Then $\chi_A(C)$ is the characteristic function of some weak random operator.

Proof. We shall show that there exists a probability space $\{\Omega, \mathfrak{S}, \mathbf{P}\}$ and a random operator $A(\omega) \in \mathbf{L}_w(\Omega, \mathbf{X})$, such that

$$\chi(C) = \mathbf{M} \exp\{i\,\mathrm{sp}\,A(\omega)\,C\}. \qquad (1.12)$$

For Ω we take the space of all real-valued functions $\psi(x, y)$, $x \in \mathbf{X}$, $y \in \mathbf{X}$, for $\mathfrak{S}$ - the σ-algebra generated by the cylindrical subsets of the afore-mentioned space. To construct a measure, we appeal to a theorem of Kolmogorov [8, p. 65]. Let $\mathbf{P}$ be the measure corresponding to the random function $\psi(x, y)$, for which

$$\mathbf{M}\exp\left\{i\sum_{k=1}^{n}\lambda_k\psi(x_k,\ y_k)\right\} = \chi\left(\sum_{k=1}^{n}\langle\lambda_k y_k\circ x_k\rangle\right), \qquad (1.13)$$

for any $x_1, \ldots, x_n, y_1, \ldots, y_n \in \mathbf{X}, \lambda_1, \ldots, \lambda_n \in \mathbf{R}$. For fixed x_k, y_k the expression on the right of (1.13) is a positive--definite function of $\lambda_1, \ldots, \lambda_n$, and so it is the characteristic function of some distribution in $\mathbf{R}^n$. Since

$$\chi\left(\sum_{k=1}^{n+1}\langle\lambda_k y_k\circ x_k\rangle\right)\Bigg|_{\lambda_{k+1}=0} = \chi\left(\sum_{k=1}^{n}\langle\lambda_k y_k\circ x_k\rangle\right),$$

it follows that the family of finite-dimensional distributions with characteristic functions (1.13) is compatible. Hence (by the above-mentioned theorem of Kolmogorov) there exists a random function $\psi(x, y)$ satisfying (1.13). For this function,

$$\begin{aligned}&\mathbf{M}\exp\{i\lambda[\psi(\alpha_1x_1+\alpha_2x_2,\ \beta_1y_1+\beta_2y_2)-\alpha_1\beta_1\psi(x_1,\ y_1)-\\&-\alpha_1\beta_2\psi(x_1,\ y_2)-\alpha_2\beta_1\psi(x_2,\ y_1)-\alpha_2\beta_2\psi(x_2,\ y_2)]\}=\\&=\chi(\lambda\langle\alpha_1x_1+\alpha_2x_2\circ\beta_1y_1+\beta_2y_2\rangle-\lambda\alpha_1\beta_1\langle x_1\circ y_1\rangle-\\&-\lambda\alpha_1\beta_2\langle x_1\circ y_2\rangle-\lambda\alpha_2\beta_1\langle x_2\circ y_1\rangle-\lambda\alpha_2\beta_2\langle x_2\circ y_2\rangle)=\\&=\chi(0)=1.\end{aligned} \qquad (1.14)$$

Therefore, for all $\alpha_1, \alpha_2, \beta_1, \beta_2 \in \mathbf{R},\ x_1, x_2, y_1, y_2 \in \mathbf{X}$

$$\begin{aligned}&\mathbf{P}\{\psi(\alpha_1x_1+\alpha_2x_2,\ \beta_1y_1+\beta_2y_2)=\alpha_1\beta_1\psi(x_1,\ y_1)+\\&+\alpha_1\beta_2\psi(x_1,\ y_2)+\alpha_2\beta_1\psi(x_2,\ y_1)+\alpha_2\beta_2\psi(x_2,\ y_2)\}=1.\end{aligned}$$

Moreover, $\psi(x, y)$ is continuous in probability at zero. But

$$\begin{aligned}&\psi(x,\ y)-\psi(x_0,\ y_0)=\psi(x-x_0,\ y)+\psi(x_0,\ y-y_0)=\\&=\psi\left(\frac{x-x_0}{|x-x_0|^{1/2}},\ |x-x_0|^{1/2}y\right)+\\&+\psi\left(|y-y_0|^{1/2}x_0,\ \frac{y-y_0}{|y-y_0|^{1/2}}\right)\end{aligned}$$

with probability 1. If $|x-x_0|\to 0$ and $|y-y_0|\to 0$, both terms on the right of the last equality approach zero in probability. Thus $\psi(x, y)$ is continuous in probability. Together with (1.14), this implies that $\psi(x, y) = (A(\omega)x, y)$, where $A(\omega) \in \mathbf{L}_w(\Omega, \mathbf{X})$. The required equality (1.12) now follows from (1.13).

2.2. Characteristic Functions of Strong and Bounded Operators

Under what conditions is $\chi_A(C)$ the characteristic function of a strong random operator? A necessary condition for this is obviously that $\chi_A(C)$ be the characteristic function of a weak random operator. In order to formulate an additional condition, to be satisfied by characteristic functions of strong random operators, we need the concept of Σ-topology. Let $\mathbf{L}_1^+(\mathbf{X})$ denote the set of all nuclear operators in $\mathbf{X}$ i.e., symmetric nonnegative operators S for which

$$\operatorname{sp} S = \sum_{k=2}^{\infty} (Se_k, e_k) < \infty.$$

The Σ-topology in $\mathbf{X}$ is defined by the neighborhoods of zero $\{\{x : (Sx, x) < 1\}; S \in \mathbf{L}_1^+(\mathbf{X})\}$. It is connected with conditions under which a function $\varphi(z)$ on $\mathbf{X}$ is the characteristic function of some $\mathbf{X}$-valued random variable $x(\omega)$, i.e., $\varphi(z) = \mathbf{M} \exp\{i(z, x(\omega))\}$. A well-known theorem of Minlos-Sazonov [15, 18] states that a necessary condition for this to be true is that $\varphi(z)$ be positive-definite and continuous in the Σ--topology at the point zero.

Theorem 2. $\chi_A(C)$ is the characteristic function of a strong random operator if and only if the following conditions hold:

(1) $\chi_A(C)$ is positive-definite, $\chi(0) = 1$ and $\chi_A(\langle x \circ y \rangle)$ is jointly continuous;

(2) $\chi_A(\langle x \circ y \rangle)$ with y held fixed is a continuous function of x at zero in the Σ-topology.

Proof. Necessity of condition (1) is obvious. Now,

$$\chi_A(\langle x \circ y \rangle) = \mathbf{M} \exp\{i(A(\omega) y, x)\},$$

and since $A(\omega)y - \mathbf{X}$ is an $\mathbf{X}$-valued r.v., the necessity of condition (2) follows from the Minlos-Sazonov theorem.

Sufficiency. Using Theorem 1 we can construct a weak random operator $A(\omega)$ with characteristic function $\chi_A(C)$. We claim that for this operator

$$\mathbf{P}\left\{\sum_{k=1}^{\infty} (A(\omega) x, e_k)^2 < \infty\right\}, \tag{1.15}$$

for any $x \in \mathbf{X}$. Since $\chi_A(\langle z \cdot x \rangle)$ is the characteristic function of $A(\omega)x$, being also a generalized random variable in $\mathbf{X}$ [12] and satsifying the conditions of the Minlos-Sazonov theorem, it follows that $A(\omega)x$ is an $\mathbf{X}$-valued r.v. and, consequently, $|A(\omega)x|$ is almost surely finite. This implies (1.15). But if $A(\omega)$ is a weak random operator satisfying (1.15) for all $x \in \mathbf{X}$ then it follows from Theorem 1 of § 1 that $A(\omega) \in \mathbf{L}_s(\Omega, \mathbf{X})$. ■

We now indicate a method for computing the distribution $|A(\omega)x|$ given a strong operator $A(\omega)$. It will clearly suffice to compute the Laplace transform

$$\psi_A(\lambda, x) = \mathbf{M} \exp\{-\lambda |A(\omega)x|^2\}, \qquad \lambda \geqslant 0.$$

The function $\psi_A(\lambda, x)$ may be defined for all $A(\omega) \in \mathbf{L}_w(\Omega, \mathbf{X})$ through the equality

$$\psi_A(\lambda, x) = \lim_{n\to\infty} \mathbf{M} \exp\left\{-\lambda \sum_{k=1}^{n} (A(\omega)x, e_k)^2\right\} \tag{1.16}$$

(the existence of the limit follows from the fact that the r.v. whose expectation is to be computed is nonnegative and decreases with n). It is readily seen that (1.15) is equivalent to the condition that $\psi_A(\lambda, x)$ be continuous in λ for $\lambda \geqslant 0$. Formula (1.16) may be written as follows:

$$\psi_A(\lambda, x) = \lim_{n\to\infty} \mathbf{M} \exp\left\{i \sqrt{2\lambda} \sum_{k=1}^{n} (A(\omega)x, e_k)\eta_k\right\}, \tag{1.17}$$

where η_k is a sequence of mutually independent Gaussian r.v.'s, independent of $A(\omega)$, such that $\mathbf{M}\eta_k = 0$, $\mathbf{D}\eta_k = 1$. But

$$\mathbf{M} \exp\left\{i \sqrt{2\lambda} \sum_{k=1}^{n} (A(\omega)x, e_k)\eta_k\right\} =$$
$$= \mathbf{M}\,\mathbf{M}\left(\exp\left\{i \sqrt{2\lambda} \sum_{k=1}^{n} (A(\omega)x, e_k)\eta_k\right\}/\eta_1, \ldots, \eta_n\right) =$$
$$= \mathbf{M}\chi_A\left(\left\langle \sqrt{2\lambda} \sum_{k=1}^{n} \eta_k e_k \circ x \right\rangle\right).$$

Let $w(\omega)$ denote a generalized Gaussian element in $\mathbf{X}$ [8, p. 409], for which $\mathbf{M} \exp\{i(z, w(\omega))\} = \exp\left\{-\frac{1}{2}|z|^2\right\}$ ($w(\omega)$ is called

Gaussian "white noise"). The r.v.'s $(w(\omega), e_k)$ are independent Gaussian, $\mathbf{M}(w(\omega), e_k) = 0$, $\mathbf{D}(w(\omega), e_k)^2 = 1$. Consequently, if P_n is the projection operator onto the subspace spanned by $\{e_1, \ldots, e_n\}$, then

$$\psi_A(\lambda, x) = \lim_{n\to\infty} \chi_A(\langle \sqrt{2\lambda}\, P_n w(\omega) \circ x\rangle).$$

Note that, thanks to the continuity of $\chi_A(\langle y \circ x\rangle)$ in the Σ-topology as a function of y, we can find a symmetric nondegenerate operator V, $\mathrm{sp}V^2 < \infty$, such that $\chi_A(\langle V^{-1} y \circ x\rangle)$ can be extended by continuity (for fixed x) to the entire space $\mathbf{X}$ [8, p. 413]. Hence follows the existence of the limit

$$\lim_{n\to\infty} \chi_A(\langle \sqrt{2\lambda}\, P_n w(\omega) \circ x\rangle) =$$
$$= \lim_{n\to\infty} \chi_A(\langle V^{-1}(\sqrt{2\lambda}\, VP_n w(\omega)) \circ x\rangle) =$$
$$= \chi_A(\langle V^{-1}(\sqrt{2\lambda}\, Vw(\omega)) \circ x\rangle) = \chi_A(\langle \sqrt{2\lambda}\, w(\omega) \circ x\rangle),$$

where $Vw(\omega)$ is a Gaussian random element in $\mathbf{X}$. Since

$$Me^{i(Vw(\omega), z)} = \exp\left\{-\frac{1}{2}|Vz|^2\right\} = \exp\left\{-\frac{1}{2}(V^2 z, z)\right\},$$

it follows that the conditions of the Minlos-Sazonov theorem are satisfied; in addition

$\mathbf{M}|VP_n w(\omega) - Vw(\omega)|^2 = \mathrm{sp}[V(P_n - I)]^2 \to 0$ as $n \to \infty$

[8, p. 406]. Obviously, $\chi_A(\langle \sqrt{2\lambda}\, w(\omega) \circ x\rangle)$ is continuous in λ. Thus the same is true of

$$\psi_A(\lambda, x) = \mathbf{M}\chi_A(\langle \sqrt{2\lambda}\, w(\omega) \circ x\rangle). \tag{1.18}$$

This formula expresses $\psi_A(\lambda, x)$ in terms of the characteristic function.

Conditions for a characteristic function $\chi_A(C)$ to be the characteristic function of a bounded random operator are not available. We shall present a sufficient condition for a random operator to be bounded, in terms of its characteristic function; this condition is also necessary and sufficient for the operator to be a Hilbert-Schmidt operator.

Theorem 3. A random operator $A(\omega)$ is a bounded Hilbert-Schmidt operator if and only if its characteristic function $\chi_A(C)$ satisfies the following conditions:

(1) For any n, there exist nonnegative Hilbert-Schmidt

operators $U_1, \ldots, U_n$, such that the function

$$\chi_A\left(\sum_{k=1}^{n} \langle U_k^{-1} x_k \circ e_k\rangle\right)$$

is continuous in $x_1, \ldots, x_n$ on $\mathbf{X}^n$;

(2) let $\{x_k(\omega),\ k = 1, 2, \ldots\}$ be independent Gaussian white-noise r.v.'s in $\mathbf{X}$; then the function

$$\lim_{n\to\infty} \mathbf{M}\chi_A\left(\lambda \sum_{k=1}^{n} \langle x_k(\omega) \circ e_k\rangle\right)$$

is continuous in λ.

Proof. Necessity of condition (1). If $A(\omega)e_k \in \mathbf{X}$, $k = 1, \ldots, n$, then

$$\chi_A\left(\sum_{k=1}^{n} \langle x_k \circ e_k\rangle\right) = \mathbf{M}\exp\left\{i \sum_{k=1}^{n} (A(\omega) e_k,\ x_k)\right\}$$

is the joint characteristic function of the r.v.'s $A(\omega)e_1, \ldots, A(\omega)e_n$. Therefore,

$$\left|\chi_A\left(\sum_{k=1}^{n} \langle x_k \circ e_k\rangle\right) - \chi_A\left(\sum_{k=1}^{n} \langle \bar{x}_k \circ e_k\rangle\right)\right| \leqslant$$

$$\leqslant \sum_{k=1}^{n} \mathbf{M}\,|\exp\{i(A(\omega) e_k,\ x_k - \bar{x}_k)\} - 1| \leqslant$$

$$\leqslant \sum_{k=1}^{n} \sqrt{2(1 - \operatorname{Re}\chi_A(\langle x_k - \bar{x}_k \circ e_k\rangle))}\,. \qquad (1.19)$$

It follows from Remark 2 in [8, p. 413] that for any k there exists a positive Hilbert-Schmidt operator U_k such that $\chi_A(\langle U_k^{-1} x \cdot e_k\rangle)$ is continuous. Hence inequality (1.19) implies the truth of condition (1). To prove the necessity of condition (2) we note that, first, as a consequence of condition (2), the expression

$$\chi_A\left(\lambda \sum_{k=1}^{n} \langle x_k(\omega) \circ e_k\rangle\right) = \chi_A\left(\lambda \sum_{k=1}^{n} \langle U_k^{-1}(U_k x_k(\omega)) \circ e_k\rangle\right)$$

is meaningful, since $U_k x_k(\omega) \in \mathbf{X}$. Second,

$$\mathbf{M}\chi_A\left(\lambda \sum_{k=1}^{n} \langle x_k(\omega) \circ e_k \rangle\right) = \mathbf{M} \exp\left\{i\lambda \sum_{k=1}^{n} (A(\omega) e_k,\ x_k(\omega))\right\}$$

(if it is assumed that $A(\omega)$ and $x_k(\omega)$ are independent). Computing the expectation in the expression on the right for fixed $A(\omega)$, we obtain

$$\mathbf{M} \exp\left\{i\lambda \sum_{k=1}^{n} (A(\omega) e_k,\ x_k(\omega))\right\} = \mathbf{M} \exp\left\{-\frac{\lambda^2}{2} \sum_{k=1}^{n} |A(\omega) e_k|^2\right\}.$$

Thus,

$$\lim_{n\to\infty} \mathbf{M} \exp\left\{-\frac{\lambda^2}{2} \sum_{k=1}^{n} |A(\omega) e_k|^2\right\} =$$
$$= \mathbf{M} \exp\left\{-\frac{\lambda^2}{2} \sum_{k=1}^{\infty} |A(\omega) e_k|^2\right\} =$$
$$= \mathbf{M} \exp\left\{-\frac{\lambda^2}{2} \operatorname{sp} A(\omega) A^*(\omega)\right\}$$

is a continuous function.

Sufficiency. It follows from condition (1) that $\chi_A(\langle x \cdot e_k \rangle)$ satisfies the conditions of the Minlos-Sazonov theorem; hence $A(\omega) e_k \in \mathbf{X}$, for all k. Further,

$$\lim_{n\to\infty} \mathbf{M}\, \chi_A\left(\lambda \sum_{k=1}^{n} \langle x_k(\omega) \circ e_k \rangle\right) =$$
$$= \lim_{n\to\infty} \mathbf{M} \exp\left\{-\frac{\lambda^2}{2} \sum_{k=1}^{n} |A(\omega) e_k|^2\right\} =$$
$$= \mathbf{M} \exp\left\{-\frac{\lambda^2}{2} \sum_{k=1}^{\infty} |A(\omega) e_k|^2\right\} = \psi(\lambda)$$

(assuming that $e^{-\infty} = 0$). In view of the continuity of this function at zero, we get

$$\mathbf{P}\left\{\sum_{k=1}^{\infty} |A(\omega) e_k|^2 = +\infty\right\} =$$
$$= \mathbf{P}\left\{1 - \exp\left\{-\frac{\lambda^2}{2} \sum_{k=1}^{\infty} |A(\omega) e_k|^2 \geqslant 1\right\}\right\} \leqslant \frac{1 - \psi(\lambda)}{1},$$

where the expression on the right tends to zero as $\lambda \to 0$ because $\psi(\lambda)$ is continuous and $\psi(0) = 1$. Thus,

$$\sum_{k=1}^{\infty} \sum_{i=1}^{\infty} (A(\omega) e_k, e_i)^2 < \infty$$

with probability 1. It remains to use Remark 2 of § 1.■

2.3. Gaussian Random Operators

A random operator $A(\omega)$ in $\mathbf{L}_w(\Omega, \mathbf{X})$ is said to be Gaussian if $(A(\omega)x, y)$ is a Gaussian random function on $\mathbf{X} \times \mathbf{X}$. Let

$$\alpha(x, y) = \mathbf{M}(A(\omega)x, y); \tag{1.20}$$

$$\beta(x, y; u, v) = \mathbf{M}(A(\omega)x, y)(A(\omega)u, v) - \alpha(x, y)\alpha(u, v)$$

(these functions are always defined). These functions uniquely determine the distribution of a random operator. In particular, the latter's characteristic function has the form

$$\chi_A\left(\sum_{k=1}^{n} \langle x_k \circ y_k \rangle\right) = \tag{1.21}$$

$$= \exp\left\{i \sum_{k=1}^{n} \alpha(y_k, x_k) - \frac{1}{2} \sum_{k,l=1}^{n} \beta(y_k, x_k, y_l, x_l)\right\}.$$

What conditions must be satisfied by the functions α and β for them to be defined by formulas (1.20) with $A(\omega)$ a Gaussian random operator?

I. $\alpha(x, y)$ is a bounded bilinear function on $\mathbf{X} \times \mathbf{X}$. Bilinearity of $\alpha(x, y)$ is obvious. That it is bounded for $|x| \leqslant 1, |y| \leqslant 1$ follows from the fact that $(A(\omega)x, y)$ is bounded in probability for $|x| \leqslant 1, |y| \leqslant 1$.

II. $\beta(x, y, u, v)$ is a bounded quadrilinear form. That it is bounded follows from the fact that $(A(\omega)x, y)$ is bounded in probability for $|x| \leqslant 1, |y| \leqslant 1$. In addition, for all $x_1, y_1, \ldots, x_n, y_n$ the form

$$\sum_{k=1}^{n} \beta(x_k, y_k; x_l, y_l)\zeta_k\zeta_l,$$

with $\zeta_1, \ldots, \zeta_n$ complex numbers, is positive semidefinite. It is easy to check that, conversely, if conditions I and II are satisfied, the characteristic function χ_A defined by (1.21) satisfies the conditions of Theorem 1; thus these conditions are sufficient for $\alpha(x, y)$ and $\beta(x, y; u, v)$ to define a Gaussian random operator (the fact that it is indeed Gaussian is readily established from (1.21)).

It turns out that the function $\alpha(x, y)$ is of the form

$$\alpha(x, y) = (Ax, y),$$

where $A \in \mathbf{L}(\mathbf{X})$. It is natural to denote $A = \mathbf{M}A(\omega)$, we shall use this notation for non-Gaussian random operators as well.

Lemma. Let $A(\omega) \in \mathbf{L}_w(\Omega, \mathbf{X})$ and suppose that $\forall x, y \in \mathbf{X}$ $\mathbf{M}(A(\omega)x, y)$. exists. Then there exists an operator in $\mathbf{L}(\mathbf{X})$, which we denote by $\mathbf{M}A(\omega)$ and call it the expectation of $A(\omega)$ such that

$$\mathbf{M}(A(\omega)x, y) = ([\mathbf{M}A(\omega)]x, y). \tag{1.22}$$

Proof. Put

$$\psi(x, y) = \mathbf{M}|(A(\omega)x, y)|.$$

This function is defined on $\mathbf{X} \times \mathbf{X}$ and, by the Fatou lemma, is lower semicontinuous:

$$\varliminf_{x\to x_0, y\to y_0} \psi(x, y) = \varliminf_{x\to x_0, y\to y_0} \mathbf{M}|(A(\omega)x, y)| \geqslant$$

$$\geqslant \mathbf{M}|(A(\omega)x_0, y_0)| = \psi(x_0, y_0),$$

since $|(A(\omega)x, y)| \xrightarrow{} |(A(\omega)x_0, y_0)|$ in probability as $x \to x_0$, $y \to y_0$. Next, $\psi(x, y)$ is semi-additive in each argument; for example:

$$\psi(x_1 + x_2, y) = \mathbf{M}|(A(\omega)x_1, y) + (A(\omega)x_2, y)| \leqslant$$

$$\leqslant \mathbf{M}|(A(\omega)x_1, y)| + \mathbf{M}|(A(\omega)x_2, y)| = \psi(x_1, y) + \psi(x_2, y).$$

Therefore $\sup\limits_{|x|\leqslant 1,\ |y|\leqslant 1} \psi(x, y) < \infty.$

Thus $\mathbf{M}(A(\omega)x, y)$ is a bounded linear form and there exists a bounded operator $\mathbf{M}A(\omega)$ satisfying (1.22).

The quadrilinear form $\beta(x, y; u, v)$ is symmetric with respect to argument pairs:

$$\beta(x, y; u, v) = \beta(u, v; x, y).$$

Define a biquadratic function by

$$\gamma(x, y) = \beta(x, y; x, y),$$

then

$$\chi_A(\langle x \circ y \rangle) = \exp\left\{i(Ay, x) - \frac{1}{2}\gamma(y, x)\right\}.$$

$A(\omega) \in \mathbf{L}_s(\Omega, \mathbf{X})$ if and only if this function is continuous in x in the Σ -topology, whatever the value of y. From Theorem 1 of [8, p. 416] it follows that this is true only if

$$\sum_{k=1}^{\infty} \gamma(y, e_k) < \infty,$$

where $\sum_{k=1}^{\infty} \gamma(y, e_k)$ is a nonnegative quadratic functional in y.

Thus,

$$\sum_{k=1}^{\infty} \gamma(y, e_k) = (By, y),$$

where B is a nonnegative symmetric operator having the following form:

$$B = \mathbf{M}A^*(\omega)A(\omega)$$

(the product is defined since $[A^*(\omega)]^* = A(\omega) \in \mathbf{L}_s(\Omega, \mathbf{X})$ Indeed,

$$\mathbf{M}(A^*(\omega)A(\omega)x, y) = \mathbf{M}\sum_k (A(\omega)x, e_k)(A(\omega)y, e_k) =$$

$$= \sum_k \beta(x, e_k; y, e_k) =$$

$$= \sum_k \frac{1}{2}[\beta(x+y, e_k; x+y, e_k) - \beta(x, e_k; x, e_k) -$$

$$- \beta(y, e_k; y, e_k)] =$$

$$= \sum_k \frac{1}{2}[\gamma(x+y, e_k) - \gamma(x, e_k) - \gamma(y, e_k)] = (Bx, y).$$

We now formulate a condition under which a Gaussian random operator is a Hilbert-Schmidt operator.

Theorem 4. Let $A(\omega) \in \mathbf{L}_s(\Omega, \mathbf{X})$ be a Gaussian random operator. It is a Hilbert-Schmidt operator if and only if

$$\mathrm{sp}\, A^*A + \mathrm{sp}\, B = \mathrm{sp}\, [\mathbf{M}A(\omega)]^*[\mathbf{M}A(\omega)] + \mathrm{sp}\, \mathbf{M}A^*(\omega)A(\omega) < \infty.$$

Proof. Sufficiency follows from the chain of equalities

$$\mathbf{M}\sum_{i,k}(A(\omega)e_i, e_k)^2 = \sum_{i,k}[\beta(e_i, e_k; e_i, e_k) + (\alpha(e_i, e_k))^2] =$$

$$= \sum_{i,k}[\gamma(e_i, e_k) + (Ae_i, e_k)^2] = \mathrm{sp}\, A^*A + \sum_i (Be_i, e_i) =$$

$$= \mathrm{sp}\, A^*A + \mathrm{sp}\, B.$$

To prove necessity, we note that the r.v.'s $\eta_{ik} = (A(\omega)e_i, e_k)$ have a Gaussian joint distribution; hence, by a lemma of [8, p. 422], the fact that the quadratic forms $\sum_{i,k\leqslant n}\eta_{ik}^2$ are bounded in probability implies that $\mathbf{M}\sum_{i,k\leqslant n}\eta_{ik}^2$ is bounded.

A few examples of Gaussian random operators follow.

(1) $\alpha(x, y) = 0$, $\beta(x, y; u, v) = (x, u)(y, v)$. The corresponding operator $A(\omega)$ is a white noise; for all x, $A(\omega)x$ is a white noise in $\mathbf{X}$; if $(x, u) = 0$, then $A(\omega)x$ and $A(\omega)u$ are independent. This operator belongs to $\mathbf{L}_w(\Omega, \mathbf{X}) \setminus \mathbf{L}_s(\Omega, \mathbf{X})$;

(2) $\alpha(x, y) = 0$, $\beta(x, y; u, v) = (x, u)(Cy, v)$, where C is a symmetric nonnegative nuclear operator:

$$A(\omega) \in \mathbf{L}_s(\Omega, \mathbf{X}),\ A^*(\omega) \in \mathbf{L}_w(\Omega, \mathbf{X}) \setminus \mathbf{L}_s(\Omega, \mathbf{X});$$

(3) $\alpha(x, y) = 0$,

$$\beta(x, y; u, v) = \sum_k (x, e_k)(y, e_k)(u, e_k)(v, e_k).$$

Then $\sum \beta(x, e_i; x, e_i) = |x|^2$, $\sum_i \beta(e_i, y; e_i, y) = |y|^2$, and so $A(\omega) \in \mathbf{L}_s(\Omega, \mathbf{X})$, $A^*(\omega) \in \mathbf{L}_s(\Omega, \mathbf{X})$. We shall show that $A(\omega) \bar{\in} \mathbf{L}(\Omega, \mathbf{X})$. For $i \neq j$,

$$\mathbf{M}(A(\omega)e_i, x)(A(\omega)e_j, y) = \sum_k (e_i, e_k)(x, e_k)(e_j, e_k)(y, e_k) = 0,$$

whatever x and y; hence $A(\omega)e_k$ for a sequence of independent Gaussian r.v.'s in $\mathbf{X}$. $(A(\omega)e_k, e_k)$ is a sequence of independent

Gaussian r.v.'s such that $\mathbf{M}(A(\omega)e_k, e_k) = 0$, $\mathbf{D}(A(\omega)e_k, e_k) = 1$.

Hence

$$\mathbf{P}\{\sup_k (A(\omega)e_k, e_k) = +\infty\} = 1$$

and $A(\omega)$ cannot be a bounded operator.

(4) An operator in $\mathbf{L}(\Omega, \mathbf{X})$ is not a Hilbert-Schmidt operator. Let $\alpha(x, y) = 0$,

$$\beta(x, y; u, v) = \sum_k \lambda_k (x, e_k)(y, e_k)(u, e_k)(v, e_k).$$

Then $\mathbf{M}(A(\omega)e_i, e_j)^2 = 0$, if $i \neq j$, $\mathbf{M}(A(\omega)e_i, e_j)^2 = \lambda_i$. Consequently, $A(\omega)$ has a diagonal form relative to the basis $\{e_k\}$ and

$$\|A(\omega)\| = \sup_i |(A(\omega)e_i, e_i)|.$$

If λ_k are chosen so that

$$\mathbf{P}\{\sup_i |(A(\omega)e_i, e_i)| > \alpha\} \leqslant$$

$$\leqslant \sum_{i=1}^{\infty} \frac{2}{\sqrt{2\pi\lambda_i}} \int_{\alpha}^{\infty} \frac{\tau}{\alpha} e^{-\frac{\tau^2}{2\lambda_i}} d\tau \leqslant \sum_{i=1}^{\infty} \frac{\sqrt{\lambda_i}}{\sqrt{2\pi}\,\alpha} e^{-\frac{\alpha^2}{2\lambda_i}} \to 0$$

as $\alpha \to \infty$, then $\mathbf{P}\{\|A(\omega)\| < \infty\} = 1$. In order that $\mathrm{sp}\, \mathbf{M}A^*(\omega)A(\omega) < \infty$, it is necessary that $\Sigma\lambda_i < \infty$.

3. Convergence of Random Operators

3.1. Weak Convergence of Random Operators

Let $A_n(\omega) \in \mathbf{L}_w(\Omega, \mathbf{X})$, $n = 0, 1, \ldots$. The sequence $A_n(\omega)$ is said to be weakly convergent to $A_0(\omega)$ if $\forall x, y \in \mathbf{X}$ the r.v.'s $(A_n(\omega)x, y)$ converge in probability to $(A_0(\omega)x, y)$.

Theorem 1. Let $A_n(\omega) \in \mathbf{L}_w(\Omega, \mathbf{X})$ and $\forall x, y \in \mathbf{X}$ the following limit exists in probability:

$$\lim_{n\to\infty} (A_n(\omega)x, y).$$

Then $A_n(\omega)$ is weakly convergent to some random operator $A_0(\omega) \in \mathbf{L}_w(\Omega, \mathbf{X})$.

Proof. Denote

$$\psi(x, y) = \lim_{n\to\infty} (A_n(\omega)x, y).$$

It is clear that $\psi(x, y)$ satisfies condition (W1) of § 1. To prove the existence of an operator $A_0(\omega)$, such that

$$\psi(x, y) = (A_0(\omega)x, y),$$

it will suffice to show that $\psi(x, y)$ is bounded in probability for $|x| \leqslant 1$, $|y| \leqslant 1$, i.e.,

$$\lim_{\alpha\to\infty} \sup_{|x|\leqslant 1, |y|\leqslant 1} \mathbf{P}\{|\psi(x, y)| > \alpha\} = 0. \tag{1.23}$$

The proof of this assertion is analogous to that of (1.9) in Lemma 2 of § 1, provided we put

$$\psi_n(x, y) = (A_n(\omega)x, y).$$

Remark. Let $\chi_{A_n}(C)$ be a sequence of characteristic functions of random operators such that for all C the limit $\lim_{n\to\infty} \chi_{A_n}(C) = \chi(C)$, exists, and for all $x, y, \in \mathbf{X}$ the function $\chi(\lambda\langle x\circ y\rangle)$ is continuous in λ. Then $\chi(C)$ is the characteristic function of some random operator. Indeed, the fact that $\chi(C)$ is positive-definite is evident. To prove that $\chi(\langle x\circ y\rangle)$ is jointly continuous in both variables, it suffices to verify that equality (1.23) holds for r.v.'s $\psi(x, y)$ (see Theorem 1 of § 2) such that

$$\mathbf{M}\exp\{i\lambda\psi(x, y)\} = \chi(\lambda\langle y\circ x\rangle) = \lim_{n\to\infty} \chi_{A_n}(\lambda\langle y\circ x\rangle) =$$

$$= \lim_{n\to\infty} \mathbf{M}\exp\{i\lambda(A_n(\omega)x, y)\}.$$

It should be noted that in establishing (1.23) (as in the case of (1.9) in Lemma 2 of § 1) one uses only the convergence of the distributions $\psi_n(x, y) = (A_n(\omega)x, y)$ to the distribution $\psi(x, y)$ (and not convergence in probability).

3.2. Strong Convergence of Random Operators

A sequence of random operators $A_n(\omega) \in \mathbf{L}_s(\Omega, \mathbf{X})$ is strongly convergent to a random operator $A_0(\omega) \in \mathbf{L}_s(\Omega, \mathbf{X})$ if

$$\forall x \in \mathbf{X},\ \varepsilon > 0 \quad \lim_{n\to\infty} \mathbf{P}\{|A_n(\omega)x - A_0(\omega)x| > \varepsilon\} = 0.$$

Theorem 2. Let $A_n(\omega) \in \mathbf{L}_s(\Omega, \mathbf{X})$ and suppose that $A_n(\omega)$ is weakly convergent to a random operator $A_0(\omega) \in \mathbf{L}_w(\Omega, \mathbf{X})$. Denote

$$\psi_{A_n}(\omega) = \mathbf{M}\exp\{-|A_n(\omega)x|^2\}.$$

If the function $\psi(x) = \inf_n \psi_{A_n}(x)$ is continuous at $x = 0$, then $A_0(\omega) \in \mathbf{L}_s(\Omega, \mathbf{X})$. In addition, if $\lim_{n\to\infty} \psi_{A_n}(x) = \psi_{A_0}(x)$, then $A_n(\omega)$ is strongly convergent to $A_0(\omega)$.

Proof. For all l and m,

$$\mathbf{M}\exp\left\{-\sum_{k=1}^{l}(A_m(\omega)x, e_k)^2\right\} \geqslant \inf_n \psi_{A_n}(x) = \psi(x).$$

It follows that

$$\mathbf{M}\exp\left\{-\sum_{k=1}^{l}(A_0(\omega)x, e_k)^2\right\} =$$

$$= \lim_{m\to\infty}\mathbf{M}\exp\left\{-\sum_{k=1}^{l}(A_m(\omega)x, e_k)^2\right\} \geqslant \psi(x);$$

$$\mathbf{P}\left\{\sum_{k=1}^{l}(A_0(\omega)x, e_k)^2 > \alpha\right\} \leqslant$$

$$\leqslant \frac{1 - \mathbf{M}\exp\left\{-\sum_{k=1}^{l}(A_0(\omega)x, e_k)^2\right\}}{1 - e^{-\alpha}} \leqslant \frac{1 - \psi(x)}{1 - e^{-\alpha}}.$$

Substituting λx for x and $\lambda^2\alpha$ for α in this inequality, we get

$$\mathbf{P}\left\{\sum_{k=1}^{l}(A_0(\omega)x, e_k)^2 > \alpha\right\} \leqslant \frac{1 - \psi(\lambda x)}{1 - e^{-\lambda^2\alpha}}.$$

Hence

$$\mathbf{P}\left\{\sum_{k=1}^{\infty}(A_0(\omega)x, e_k)^2 = +\infty\right\} \leqslant 1 - \psi(\lambda x),$$

for any $\lambda > 0$. Letting $\lambda \to 0$ $(\psi(0) = 1)$ and using Theorem 2 of § 1, we find that $A_0(\omega) \in \mathbf{L}_s(\Omega, \mathbf{X})$. For $n = 0, 1, ..$, we now put

$$\zeta_n = 1 - \exp\{-|A_n(\omega) x|^2\}.$$

We may assume without loss of generality that $(A_n(\omega) x, e_k) \to \to (A_0(\omega) x, e_k)$ for all k, with probability 1. Then

$$\varliminf_{n\to\infty} \zeta_n \geqslant \varliminf_{n\to\infty} \left(1 - \exp\left\{-\sum_{k=1}^{l} (A_n(\omega) x, e_k)^2\right\}\right) =$$

$$= 1 - \exp\left\{-\sum_{k=1}^{l} (A_0(\omega) x, e_k)^2\right\}$$

with probability 1, and so

$$\mathbf{P}\{\varliminf_{n\to\infty} \zeta_n \geqslant \zeta_0\} = 1. \tag{1.24}$$

Denote $\bar{\zeta}_n = \zeta_n \vee \zeta_0$. Then $\mathbf{P}\{\bar{\zeta}_n - \zeta_0 < 0\} = 0$, and it follows from (1.24) that for $\varepsilon > 0$

$$\lim_{n\to\infty} \mathbf{P}\{\bar{\zeta}_n - \zeta_n > \varepsilon\} = \lim_{n\to\infty} \mathbf{P}\{\zeta_0 - \zeta_n > \varepsilon\} = 0.$$

Therefore, $\bar{\zeta}_n - \zeta_n \to 0$ in probability and

$$\mathbf{M}(\bar{\zeta}_n - \zeta_n) \to 0.$$

Moreover, by the assumptions of our theorem $\mathbf{M}\zeta_n \to \mathbf{M}\zeta_0$ and so

$$\mathbf{M}|\bar{\zeta}_n - \zeta_0| \to 0.$$

Since $\bar{\zeta}_n \geqslant \zeta_n$, it follows from this that $\bar{\zeta}_n \to \zeta_0$ and $\zeta_n \to \zeta_0$ in probability. In order to show that $A_n(\omega)x \to A_0(\omega)x$ in probability, it will suffice to prove that from any sequence of natural numbers n_k one can extract a subsequence n'_k, such that $A_{n'_k}(\omega) x \to A_0(\omega) x$ with probability 1. In fact, this will be the case for any subsequence n'_k, such that

$$\mathbf{P}\{(A_{n'_k}(\omega) x, e_i) \to (A_0(\omega) x, e_i)\} = 1 \quad \forall i;$$

$$\mathbf{P}\{|A_{n'_k}(\omega) x| \to |A_0(\omega) x|\} = 1,$$

since the convergence of (y_n, e_i) to (y_0, e_i) and the fact that $|y_n| \to |y_0|$ imply the convergence of y_n to y_0

3.3. Convergence of Distributions corresponding to Random Operators

The distributions of random operators $A_n(\omega) \in \mathbf{L}_w(\Omega, \mathbf{X})$ are weakly convergent to the distribution of a random operator $A_0(\omega)$ if, for any $m, x_1, \ldots, x_m, y_1, \ldots, y_m \in \mathbf{X}$, the joint distribution of the r.v.'s $\{(A_0(\omega) x_k, y_k), k = 1, \ldots, m\}$. converges to the joint distribution of the r.v.'s $\{(A_n(\omega) x_k, y_k), k = 1, \ldots, m\}$

The distributions of random operators $A_n(\omega) \in \mathbf{L}_s(\Omega, \mathbf{X})$ are strongly convergent to the distribution of a random operator $A_0(\omega) \in \mathbf{L}_s(\Omega, \mathbf{X})$, if, for all $m, x_1, \ldots, x_m \in \mathbf{X}$ the joint distribution of the $\mathbf{X}$-valued r.v.'s $\{A_n(\omega) x_k, k = 1, \ldots, m\}$ converges to the joint distribution of the r.v.'s $\{A_0(\omega) x_k, k = 1, \ldots, m\}$. The latter assertion is equivalent to the equality

$$\lim_{n\to\infty} \mathbf{M} f(A_n(\omega) x_1, \ldots, A_n(\omega) x_m) =$$

$$= \mathbf{M} f(A_0(\omega) x_1, \ldots, A_0(\omega) x_m)$$

for any bounded continuous function $f(x_1, \ldots, x_m)$ on $\mathbf{X}^m$.

It is readily seen that weak (strong) convergence of random operators implies weak (strong) convergence of their distributions.

A. The distributions of $A_n(\omega)$ are weakly convergent to the distribution of $A_0(\omega)$ if and only if

$$\lim_{n\to\infty} \chi_{A_n}(C) = \chi_{A_0}(C) \quad \forall C \in L_0, \tag{1.25}$$

where $\chi_{A_n}(\cdot)$ is the characteristic function of $A_n(\omega)$.

B. Suppose that, in addition to (1.25), the following conditions hold:

(1) $A_n(\omega) \in \mathbf{L}_s(\Omega, \mathbf{X}),\ n = 0, 1, \ldots;$

(2) $\lim_{n\to\infty} \mathbf{M} \exp\{-|A_n(\omega) x|^2\} = \mathbf{M} \exp\{-|A_0(\omega) x|^2\}\ \forall x \in \mathbf{X}.$

Then the distributions of $A_n(\omega)$ are strongly convergent to that of $A_0(\omega)$.

Proposition B is a corollary of the next theorem and Theorem 2.

Theorem 3. Let $A_n(\omega)$ be a sequence of random operators, $\chi_{A_n}(\cdot)$ their characteristic functions; suppose that the limit

$$\chi(C) = \lim_{n\to\infty} \chi_{A_n}(C) \quad \forall C \in \mathbf{L}_0$$

exists and that the function $\chi(\lambda\langle x \circ y\rangle)$ is continuous in λ. Then one can construct a sequence of random operators $\{\hat{A}_n(\omega), n = 0, 1, \ldots\}$, such that $\chi_{\hat{A}_n}(\cdot) = \chi_{A_n}(\cdot)$, $n = 1, 2, \ldots, \chi_{\hat{A}_0}(\cdot) = \chi(\cdot)$ and $A_n(\omega)$ is weakly convergent to $\hat{A}_0(\omega)$.

Proof. It follows from the Remark to Theorem 1 that $\chi(C)$ is the characteristic function of some random operator $A_0(\omega)$. Choose some dense countable set $\{x_k, k = 1, 2, \ldots\}$. Since for all m the joint distribution of the r.v.'s $\{(A_n(\omega)x_i, x_j), i, j = 1, \ldots, m\}$ converges to the joint distribution of the r.v.'s $\{(A_0(\omega)x_i, x_j), i, j = 1, \ldots, m\}$ (see [20], § 6, ch. 1), it follows that we can construct random variables $\{\beta_n(x_i, x_j), i, j = 1, 2, \ldots\}$ on some probability space, such that:

(1) for all $n = 0, 1, \ldots$ the joint distribution of $\{\beta_n(x_i, x_j), i, j = 1, 2, \ldots\}$ is identical to the joint distribution of the r.v.'s $\{(A_n(\omega)x_i, x_j); i, j = 1, 2, \ldots\}$;

(2) $\forall i,j$ $\beta_n(x_i, x_j)$ converges in probability to $\beta_0(x_i, x_j)$. Now, for all x and y, the following limit exists in probability:

$$(A_n(\omega)x, y) = \lim_{x_{i_m}\to x, x_{j_m}\to y} (A_n(\omega)x_{i_m}, x_{j_m}),$$

and so condition (1) implies the existence in probability of the limit

$$\lim_{x_{i_m}\to x, x_{j_m}\to y} \beta_n(x_{i_m}, x_{j_m}),$$

which we denote by $\beta_n(x, y)$. It is easy to verify that the joint distribution of the r.v.'s $\{\beta_n(\bar{x}_j, \bar{y}_j), j = 1, \ldots, m\}$ is identical to that of the r.v.'s $\{(A_n(\omega)\bar{x}_j, \bar{y}_j), j = 1, \ldots, m\}$. Consequently,

$$\beta_n(x, y) = (B_n(\omega)x, y),$$

where $B_n(\omega)$ are certain random operators. We claim that $B_n(\omega)$ is weakly convergent to $B_0(\omega)$. To prove this, we first show that

$$\lim_{\alpha\to\infty} \sup_{|x|\leqslant 1, |y|\leqslant 1} \sup_n \mathbf{P}\{|(B_n(\omega)x, y)| > \alpha\} = 0. \qquad (1.26)$$

The proof of this relationship is analogous to the proof of (1.9) in Lemma 2 of § 1 (the role of $\mathbf{P}\{|\psi(x, y)| > \alpha\}$ is now taken by $\sup_n \mathbf{P}\{|(B_n(\omega)x, y)| > \alpha\}$). Now for any fixed n,

$$\lim_{\alpha\to\infty} \sup_{x\in S_\delta(x_1)} \sup_{y\in S_\delta(y_1)} \mathbf{P}\{|(B_n(\omega)x, y)| > \alpha\} = 0,$$

and so, if (1.26) were false, there would exist $\varepsilon > 0$, sequences $n_k \to \infty$, $\delta_k \to 0$, $\alpha_k \to \infty$, and nested spheres $S_{\delta_k}(x_k) \supset S_{\delta_{k+1}}(x_{k+1})$, $S_{\delta_k}(y_k) \supset S_{\delta_{k+1}}(y_{k+1})$, such that

$$\inf_{x\in S_{\delta_k}(x_k)} \inf_{y\in S_{\delta_k}(y_k)} \mathbf{P}\{|(B_{n_k}(\omega)x, y)| > \alpha_k\} \geqslant \varepsilon.$$

This contradicts the fact that the distributions $(B_{n_k}(\omega)x_0, y_0)$ converge to $(B_0(\omega)x_0, y_0)$, where $x_0 \in \bigcap_k S_{\delta_k}(x_k)$, $y_0 \in \bigcap_k S_{\delta_k}(y_k)$. Thus (1.26) is necessarily true.

Now let $x_{i_m} \to x$, $x_{j_m} \to y$. Then

$$\overline{\lim_{n\to\infty}} \mathbf{P}\{|(B_n(\omega)x, y) - (B_0(\omega)x, y)| > \delta\} \leqslant$$

$$\leqslant \overline{\lim_{n\to\infty}} \left[\mathbf{P}\left\{|(B_n(\omega)x, y) - (B_n(\omega)x_{i_m}, x_{j_m})| > \frac{\delta}{3}\right\}\right] +$$

$$+ \mathbf{P}\left\{|(B_0(\omega)x, y) - (B_0(\omega)x_{i_m}, x_{j_m})| > \frac{\delta}{3}\right\} +$$

$$+ \mathbf{P}\left\{|(B_n(\omega)x_{i_m}, x_{j_m}) - (B_0(\omega)x_{i_m}, x_{j_m})| > \frac{\delta}{3}\right\} \leqslant$$

$$\leqslant 2 \sup_n \left[\mathbf{P}\left\{|(B_n(\omega)(x - x_{i_m}), y)| > \frac{\delta}{6}\right\} +\right.$$

$$\left. + \mathbf{P}\left\{|(B_n(\omega)x_{i_m}, y - x_{j_m})| > \frac{\delta}{6}\right\}\right] \leqslant$$

$$\leqslant 4 \sup_n \sup_{|x'|\leqslant 1} \sup_{|y'|\leqslant 1} \mathbf{P}\left\{|B_n(\omega)x', y')| >\right.$$

$$\left. > \frac{\delta}{(|y| + |x_{i_m}|)(|x - x_{i_m}| + |y - y_{j_m}|)}\right\}.$$

It follows from (1.26) that the last expression approaches zero as $x - x_{i_m} \to 0$, $y - y_{j_m} \to 0$. ■

CHAPTER 2

FUNCTIONS OF RANDOM OPERATORS

4. Spectral Representations for Symmetric Random Operators

4.1. Symmetric Random Operators and Selfadjoint Extensions

A random operator $A(\omega) \in \mathbf{L}_w(\Omega, \mathbf{X})$ is said to be symmetric if $\mathbf{P}\{(A(\omega)x, y) = (x, A(\omega)y)\} = 1$. A symmetric operator may belong to $\mathbf{L}_s(\Omega, \mathbf{X})$ or to $\mathbf{L}(\Omega, \mathbf{X})$.

Consider a symmetric operator in $\mathbf{L}_w(\Omega, \mathbf{X})$, for which there exists a basis $\{e_k\}$ in $\mathbf{X}$ such that for all i, with probability 1,

$$\sum_{k=1}^{\infty} (A(\omega)e_i, e_k)^2 < \infty. \tag{2.1}$$

For this operator the elements $A(\omega)e_i$, $i = 1, 2, \ldots$ are defined.

Choose numbers $\sigma_k \downarrow 0$ in such a way that the series

$$\sum_{k,i} (A(\omega)e_k, A(\omega)e_i)^2 \sigma_k^2 \sigma_i^2 = \rho^2(\omega).$$

is convergent with probability 1. Let B be a symmetric linear operator such that $Be_k = \sigma_k e_k$. Then for any x, which is a linear combination of the $\{e_k\}$, we have

$$|A(\omega)Bx|^2 = \sum_{k,i} (A(\omega)Be_k, A(\omega)Be_i)(x, e_k)(x, e_i) =$$

$$= \sum_{k,i} (A(\omega)e_k, A(\omega)e_i)\sigma_k\sigma_i(x, e_k)(x, e_i) \leqslant$$

$$\leqslant \sqrt{\sum_{k,i} (A(\omega)e_k, A(\omega)e_i)^2 \sigma_k^2\sigma_i^2} \sqrt{\sum_{k,i} (x, e_k)^2 (x, e_i)^2} = \rho(\omega)|x|^2.$$

Therefore $A(\omega)B \in \mathbf{L}(\Omega, \mathbf{X})$. Let $\mathbf{D}$ denote the range of B; it is dense in $\mathbf{X}$. Consider the operator

$$C(\omega) = BA(\omega)B,$$

which is a symmetric operator whose range is a subset of **D**. Put

$$A'(\omega) = B^{-1}C(\omega)B^{-1}.$$

For any ω, this operator is an unbounded symmetric linear operator with everywhere dense domain **D** (independent of ω). Given $x, y \in \mathbf{D}$ we have, with probability 1,

$$(A'(\omega)x, y) = (A(\omega)x, y).$$

Since a symmetric operator has a closed extension (such an extension for the operator A is A^*), it clearly has an extension which is a symmetric closed operator.

We shall construct a selfadjoint extension for a closed symmetric operator. To this end we need the complex extension of the original Hilbert space. Let $\tilde{\mathbf{X}}$ denote the collection of elements $x + iy$, where $x, y \in \mathbf{X}$. Addition of elements $\tilde{\mathbf{X}}$ and multiplication by complex numbers are defined in the natural way, making $\tilde{\mathbf{X}}$ a linear space. We define a scalar product in $\tilde{\mathbf{X}}$ by $(x + iy, u + iv) = (x, y) + (y, v) + i(y, u) - i(x, v)$, thus converting $\tilde{\mathbf{X}}$ into a complex Hilbert space. We consider **X** as a subspace of $\tilde{\mathbf{X}}$. Any basis $\{e_k\}$ in **X** is at the same time a basis in $\tilde{\mathbf{X}}$. If $z = x + iy \in \tilde{\mathbf{X}}$, we denote $\bar{z} = x - iy$.

Linear operators defined in **X**, extend by linearity to $\tilde{\mathbf{X}}$. If an operator A is defined on **D**, it is defined in $\tilde{\mathbf{X}}$ on the domain $\tilde{\mathbf{D}}$, $\tilde{\mathbf{D}} = \{x + iy,\ x \in D,\ y \in D\}$ by the equality $A(x + iy) = Ax + iAy$ If A is symmetric in **X**, it is also symmetric in $\tilde{\mathbf{X}}$.

Lemma. Let A be a closed symmetric operator defined in a domain $\mathbf{D}_A$ which is dense in $\tilde{\mathbf{X}}$ and together with any $x = x_1 + ix_2$ also contains $\operatorname{Re} x = x_1$. If $\operatorname{Im}(Ax, y) = 0$ whenever $\operatorname{Im} x = 0$, $\operatorname{Im} y = 0$, then A has a selfadjoint extension.

Proof. Put

$$\mathbf{L}_\pm = \{y\colon\ y = Ax \pm ix\},$$

$\mathbf{L}_\pm$ are linear subspaces. Since

$$(Ax \pm ix, Ax \pm ix) = (Ax, Ax) + (x, x),$$

it follows that, if $z_n = Ax_n \pm ix_n$ converges to come limit, then x_n and Ax_n both converge to limits; hence, since the operator is closed, it follows that $z = \lim z_n = Ax_0 \pm ix_0$, and so $\mathbf{L}_\pm$ are closed linear subspaces. Let $\mathbf{N}_\pm$ denote the orthogonal complements of $\mathbf{L}_\pm$. If $y \in \mathbf{N}_+$, then

$$(Ax + ix, y) = 0, \quad \forall x \in \mathbf{D}_A.$$

Thus

$$(Ax, y) = -i(x, y), \quad y \in \mathbf{D}_{A^*};$$
$$(x, A^*y) = -i(x, y) = (x, iy).$$

Thus, A^* is defined on $\mathbf{N}_+$ and $A^*y = iy$. Similarly, A^* is defined on N_- and $A^*y = -iy$. We shall show that if $y \in \mathbf{N}_+$, then $\bar{y} \in N_-$. Suppose that $\operatorname{Im} x = 0$. Then, putting $y = y_1 + iy_2$, we obtain

$$\begin{aligned}(Ax - ix, \bar{y}) &= (Ax - ix, y_1 - iy_2) = \\ &= (Ax, y_1) + (x, y_2) - i(x, y_2) + i(Ax, y_2) = \\ &= \overline{(Ax, y_1) + (x, y_2) + i(x, y_1) - i(Ax, y_2)} = \\ &= \overline{(Ax + ix, y_1 + iy_2)} = \overline{(Ax + ix, y)} = 0\end{aligned}$$

(by assumption $y \in \mathbf{N}_+$). Since any $x \in \mathbf{D}_A$ is of the form $x_1 + ix_2$, where $x_1, x_2 \in \mathbf{D}_A$, $\operatorname{Im} x_1 = \operatorname{Im} x_2 = 0$, it follows that $(Ax - ix, \bar{y}) = 0 \ \forall x \in \mathbf{D}_A$. In exactly the same way one shows that if $y' \in N_-$, then $\bar{y}' \in \mathbf{N}_+$. The mapping

$$Ty = \bar{y}$$

is isometric from $\mathbf{N}_+$ onto $\mathbf{N}_-$, since $(y, y) = (\bar{y}, \bar{y})$. Let $\mathbf{N}$ denote the set of elements expressible as $y + i\bar{y}$, where $y \in \mathbf{N}_+$. The operator A^* is defined on $\mathbf{N}$ and $A^*(y + i\bar{y}) = iy + \bar{y}$. Consider the operator $\bar{A}$, defined on $\mathbf{D}_A + \mathbf{N}$ by $\bar{A}(x + y) = Ax + A^*y$ for $x \in \mathbf{D}_A$, $y \in \mathbf{N}$ (this definition is self-consistent, because $A = A^*$ on $\mathbf{D}_A$). We claim that $\bar{A}$ is symmetric. Indeed, by assumption $(\bar{A}x, y) = (x, \bar{A}y)$ for $x, y \in \mathbf{D}_A$, and by definition $(\bar{A}x, y) = (x, A^*y) = (x, \bar{A}y)$ for $x \in \mathbf{D}_A$, $y \in \mathbf{N}$. Let us check the symmetry condition for $y_1, y_2 \in \mathbf{N}$. Let $y_k = u_k + i\bar{u}_k$. Then

$$\begin{aligned}(\bar{A}y_1, y_2) &= (\bar{A}(u_1 + i\bar{u}_1), u_2 + i\bar{u}_2) = (iu_1 + \bar{u}_1, u_2 + i\bar{u}_2) = \\ &= i(u_1, u_2) - i(\bar{u}_1, \bar{u}_2) + (u_1, \bar{u}_2) + (\bar{u}_1, u_2) = \\ &= (u_1, \bar{u}_2) + (\bar{u}_1, u_2)\end{aligned}$$

in view of the isometry of the mapping $T(u_1, u_2) = (\bar{u}_1, \bar{u}_2)$. Similarly,

$$(y_1, \bar{A}y_2) = (u_1 + i\bar{u}_1, iu_2 + \bar{u}_2) = (u_1, \bar{u}_2) + (\bar{u}_1, u_2).$$

Thus $\bar{A}$ is symmetric. We now show that $\bar{A}$ is closed. Let $z_n = x_n + y_n + i\bar{y}_n$, where $x_n \in \mathbf{D}_A$, $y_n \in \mathbf{N}_+$. Then $\bar{A}z_n = Ax_n + iy_n + \bar{y}_n$. If the sequence z_n and $\bar{A}z_n$, possess limits, then the same is true of $(\bar{A}z_n + iz_n) =$ $= Ax_n + ix_n + 2iy_n$. Since $Ax_n + ix_n \in \mathbf{L}_+$, $2iy_n \in \mathbf{N}_+$, the limits $\lim y_n$ and $\lim (Ax_n + ix_n)$ exist. Consequently, so do $\lim x_n$ and $\lim Ax_n$. If $\lim x_n = x$, then $x \in \mathbf{D}_A$ and $Ax_n \to Ax$, $\lim y_n = y \in \mathbf{L}_+$. Thus $z = \lim z_n = x + y + i\bar{y} \in \mathbf{D}_A + \mathbf{N}$ and $\bar{A}z = \lim \bar{A}z_n$. We have thus shown that $\bar{A}$ is closed. For A to be selfadjoint it is sufficient that $\bar{\mathbf{L}}_\pm = \{y\colon\ y = \bar{A}x \pm \pm ix\} = \tilde{\mathbf{X}}$. Let $z \perp \mathbf{L}_+$ say. Then $z \in \mathbf{N}_+$, since $\bar{\mathbf{L}}_+ \supset \mathbf{L}_+$. If $x = z + i\bar{z}$, then $x \in N$ and $0 = (\bar{A}x + ix, z) =$ $= (iz + \bar{z} + iz - \bar{z}, z) = 2i\,(z, z)$, $z = 0$. Thus $\bar{A}$ is indeed selfadjoint.

Consider the set of operators $A'(\omega) = B^{-1}C(\omega)B^{-1}$ and let $\bar{A}(\omega)$ denote their selfadjoint extensions, constructed as just shown. Clearly, for $x, y \in \mathbf{D}$ with probability 1,

$$(\bar{A}(\omega)x,\ y) = (A(\omega)x,\ y).$$

Any operator in the family $\{\bar{A}(\omega)\}$ has its own domain $\mathbf{D}_\omega \in \mathbf{X}$, on which it is selfadjoint; the join of all these domains, $\bigcap_\omega \mathbf{D}_\omega \supset \mathbf{D}$ is dense in $\mathbf{X}$. An operator $\bar{A}(\omega)$, possessing this property is called a selfadjoint random operator if it is stochastically equivalent on $\mathbf{D}$ to some weak random operator. This subsection may therefore be summarized as follows: given a symmetric random operator satisfying condition (2.1), we have constructed a selfadjoint random extension.

4.2. Spectral Representation of a Selfadjoint Random Operator

Let $\bar{A}(\omega)$ be a selfadjoint random operator. Using the spectral theory of selfadjoint operators ([1], ch. 6), one sees that for any ω there exists a family of projections $\{E_\lambda(\omega), \lambda \in \mathbf{R}\}$ such that

(1) $E_\lambda(\omega)E_\mu(\omega) = E_\lambda(\omega)$ whenever $\lambda < \mu$;

(2) $(E_\lambda(\omega)x, x)$ is a nondecreasing left-continuous function for $x \in \mathbf{X}$

$$\lim_{\lambda\to-\infty}(E_\lambda(\omega)x, x) = 0, \quad \lim_{\lambda\to\infty}(E_\lambda(\omega)x, x) = (x, x);$$

(3) $E_\lambda(\omega)$ commutes with $\bar{A}(\omega)$;

(4) for $-\infty \leqslant \lambda < \mu \leqslant +\infty$ and $x \in \mathbf{D}_\omega$,

$$|\bar{A}(\omega)(E_\mu(\omega) - E_\lambda(\omega))x|^2 = \int_\lambda^\mu \sigma^2 d(E_\sigma(\omega)x, x); \qquad (2.2)$$

$$(\bar{A}(\omega)(E_\mu(\omega) - E_\lambda(\omega))x, x) = \int_\lambda^\mu \sigma d(E_\sigma(\omega)x, x); \qquad (2.3)$$

(5) if $-\infty < \lambda < \mu < +\infty$, then (2.2) and (2.3) are valid for all $x \in \mathbf{X}$. We shall show that $E_\lambda(\omega) \in \mathbf{L}(\Omega, \mathbf{X})$. Let us recall the construction of the family $\{E_\lambda(\omega), \lambda \in \mathbf{R}\}$. Define an operator $U(\omega)$ by

$$U(\omega) = (iI - \bar{A}(\omega))(iI + \bar{A}(\omega))^{-1}$$

for all $x \in \tilde{\mathbf{X}}$. $U(\omega)$ is a unitary operator:

$$(U(\omega)x, y) = ((iI - \bar{A}(\omega))(iI + \bar{A}(\omega))^{-1}x, y) =$$
$$= ((iI + \bar{A}(\omega))^{-1}x, (iI + \bar{A}(\omega))y).$$

For $y \in \mathbf{D}$, we have $(iI + \bar{A}(\omega))y = iy + A(\omega)y$ with probability 1, so $(iI + \bar{A}(\omega))y$ is $\mathfrak{S}$-measurable. We claim that $(iI + A(\omega))^{-1}$ is also $\mathfrak{S}$-measurable. Let $P(\omega)$ denote the projection onto $\mathbf{N}_+(\omega) = \{y \in \mathbf{X}: (A(\omega)x + ix, y) = 0, \forall x \in \mathbf{D}\}$. Since

$$\mathbf{N}_+(\omega) = \bigcap_k \{y \in \tilde{\mathbf{X}}: (A(\omega)e_k + ie_k, y) = 0\},$$

it follows that $\mathbf{N}_+(\omega)$ is the set of zeros of the expression

$$\sum_{k=1}^\infty |(A(\omega)e_k + ie_k, y)|^2 \sigma_k^2,$$

where the σ_k are so chosen that

$$\sum_{k=1}^\infty \sigma_k^2(|A(\omega)e_k|^2 + 1) < \infty.$$

Then

$$\sum_{k=1}^\infty |(A(\omega)e_k + ie_k, y)|^2 \sigma_k^2 = (V(\omega)y, y),$$

where $V(\omega)$ is a nonnegative symmetric operator in $\mathbf{L}(\Omega, \mathbf{X})$. Since

$$\operatorname{sp} V(\omega) = \sum_{k=1}^{\infty} \sum_{j=1}^{\infty} |(A(\omega) e_k + ie_k, e_j)|^2 \sigma_k^2 =$$

$$= \sum_{k=1}^{\infty} (|A(\omega) e_k|^2 + 1) \sigma_k^2 < \infty,$$

it follows that $V(\omega)$ is absolutely continuous for all ω. It is readily seen that for all x

$$P(\omega) x = \lim_{t \to \infty} e^{-tV(\omega)} x.$$

Thus $P(\omega)$ is an $\mathfrak{S}$-measurable operator. Let $z(\omega) = (iI + \bar{A}(\omega))^{-1} x$. Then $z(\omega) = z_1(\omega) + z_2(\omega) + iz_2(\omega)$, where $z_1(\omega) \in \mathbf{D}_\omega$, $z_2(\omega) \in \mathbf{N}_+(\omega)$ ($\mathbf{D}_\omega$ is the domain of the closure $A'(\omega)$), and

$$x = (iI + \bar{A}(\omega))(z_1(\omega) + z_2(\omega) + i\bar{z}_2(\omega)) =$$

$$= \bar{A}(\omega) z_1(\omega) + iz_1(\omega) + 2iz_2(\omega).$$

At the same time, $\bar{A}(\omega) z_1(\omega) + iz_1(\omega) \in \mathbf{L}_+(\omega)$. Therefore $z_2(\omega) = \frac{1}{2i} P(\omega) x$. We thus obtain the following equation for $z_1(\omega)$

$$\bar{A}(\omega) z_1(\omega) + iz_1(\omega) = x - P(\omega) x.$$

Let $\{x_k, k \geqslant 1\}$ be a dense subset of $\mathbf{D}$, $y_k(\omega) = A(\omega) x_k + ix_k$, each $y_k(\omega)$ is $\mathfrak{S}$-measurable. The set $\{y_k(\omega), k \geqslant 1\}$ is dense in $\mathbf{L}_+(\omega)$. As $\|(\bar{A}(\omega) + iI)^{-1}\| \leqslant 1$ on $\mathbf{L}_+(\omega)$, we have

$$|y_k(\omega) - (x - P(\omega) x)| \geqslant |x_k - z_1(\omega)|. \tag{2.4}$$

Let $z_1^{(n)}(\omega) = x_k$, if $|y_i(\omega) - (x - P(\omega) x)| > \frac{1}{n}$ for $i < k$ and $|y_k(\omega) - (x - P(\omega) x)| \leqslant \frac{1}{n}$. Then $|z^{(n)}(\omega) - z_1(\omega)| =$

$|z_1(\omega) - x_k| \leqslant \frac{1}{n}$ by virtue of (2.4). Obviously, $z^{(n)}(\omega)$ is $\mathfrak{S}$-measurable, hence the same is true of $z_1(\omega)$ as the limit of $z^{(n)}(\omega)$. Thus, $(iI + A(\omega))^{-1} x$ is $\mathfrak{S}$-measurable, and therefore $(U(\omega) x, y)$ is also $\mathfrak{S}$-measurable for $x \in \tilde{\mathbf{X}}$, $y \in \mathbf{D}$. Using the boundedness of $U(\omega)$ ($\|U(\omega)\| = 1$) and the denseness of $\mathbf{D}$ in $\tilde{\mathbf{X}}$, we see that $(U(\omega)x, y)$ is $\mathfrak{S}$-measurable for all $x, y \in \mathbf{X}$. Therefore $U(\omega) \in \mathbf{L}(\Omega, \mathbf{X})$. Thus the power $U^k(\omega)$ belongs to $\mathbf{L}(\Omega, \mathbf{X})$ for all integers k. Since the operators $E_\lambda(\omega)$ are defined as limits

$$(E_\lambda(\omega)x, y) = \lim_{n\to\infty} \sum_{k=-\infty}^{\infty} \alpha_k^{(n)}(\lambda)(U^k(\omega)x, y),$$

where $\alpha_k^{(n)}(\lambda)$ are independent of ω, it follows that $(E_\lambda(\omega)x, y)$ is $\mathfrak{S}$-measurable, and so $E_\lambda(\omega) \in \mathbf{L}(\Omega, \mathbf{X})$.

4.3. Spectral Representation of a Strong Symmetric Operator

Suppose that $A(\omega) \in \mathbf{L}_s(\Omega, \mathbf{X})$ and that $A(\omega)$ is symmetric. Then $A(\omega)e_k$ is defined for any basis $\{e_k\}$. Therefore, in view of what we have proved, we can find a dense linear subset in $\mathbf{X}$, say $\mathbf{D}$, and a family of projections $\{E_\lambda(\omega), \lambda \in \mathbf{R}\}$, all elements of $\mathbf{L}(\Omega, \mathbf{X})$, which satisfy conditions (1), (2) and, with probability 1, the following conditions:

(3') for $\lambda < \mu$ and $x \in \mathbf{D}$

$$(E_\mu(\omega) - E_\lambda(\omega))A(\omega)x = A(\omega)(E_\mu(\omega) - E_\lambda(\omega))x; \quad (2.5)$$

(4') for $-\infty \leqslant \lambda < \mu \leqslant +\infty$, $x \in \mathbf{D}$

$$|A(\omega)(E_\mu(\omega) - E_\lambda(\omega))x|^2 = \int_\lambda^\mu \sigma^2 d(E_\sigma(\omega)x, x); \quad (2.6)$$

$$(A(\omega)(E_\mu(\omega) - E_\lambda(\omega))x, x) = \int_\lambda^\mu \sigma d(E_\sigma(\omega)x, x) \quad (2.7)$$

(where the integration is over the set $[\lambda, \mu)$). We first prove (2.5). Since $A(\omega)x$ is an element of $\mathbf{X}$ and $E_\mu(\omega) - E_\lambda(\omega)$ is a bounded operator, the left-hand side of (2.5) is meaningful. The operator $A(\omega)(E_\mu(\omega) - E_\lambda(\omega))$ should be interpreted as a weak operator:

$$(A(\omega)(E_\mu(\omega) - E_\lambda(\omega))x, y) = ((E_\mu(\omega) - E_\lambda(\omega))x, A(\omega)y)$$

(the expression on the right of this equality is meaningful). Choose a basis in $\mathbf{D}$, say $\{e_k\}$. We use the fact that $\bar{A}(\omega)x = A(\omega)x$ on $\mathbf{D}$ with probability 1. Letting $\bar{A}(\omega)$ denote the selfadjoint extension of $A(\omega)$ in $\mathbf{D}$, we obtain

$$((E_\mu(\omega) - E_\lambda(\omega))A(\omega)x, e_k) =$$
$$= ((E_\mu(\omega) - E_\lambda(\omega))\bar{A}(\omega)x, e_k) = (\bar{A}(\omega)(E_\mu(\omega) -$$

$$- E_\lambda(\omega))\, x,\, e_k) = ((E_\mu(\omega) - E_\lambda(\omega))\, x,\, \bar{A}(\omega)\, e_k) =$$
$$= ((E_\mu(\omega) - E_\lambda(\omega))\, x,\, A(\omega)\, e_k) =$$
$$= (A(\omega)(E_\mu(\omega) - E_\lambda(\omega))\, x,\, e_k). \tag{2.8}$$

Therefore,

$$\sum_{k=1}^{\infty} (A(\omega)(E_\mu(\omega) - E_\lambda(\omega))\, x,\, e_k)^2 =$$

$$= \sum_{k=1}^{\infty} ((E_\mu(\omega) - E_\lambda(\omega))\, A(\omega)\, x,\, e_k)^2 =$$

$$= |(E_\mu(\omega) - E_\lambda(\omega))\, A(\omega)\, x|^2 \leqslant |A(\omega)\, x|^2 < \infty$$

and $A(\omega)(E_\mu(\omega) - E_\lambda(\omega))x$ takes values in $\mathbf{X}$. The required equality (2.5) follows from (2.8). Equalities (2.6), (2.7) are consequences of (2.5), (2.2), (2.3), and of the fact that $A(\omega)$ coincides with $\bar{A}(\omega)$ on $\mathbf{D}$ with probability 1.

We now show that formulas (2.5)-(2.7) may be extended to all $x \in \mathbf{X}$. Since $(E_\mu(\omega) - E_\lambda(\omega))\, A(\omega)$ and $A(\omega)(E_\mu(\omega) - E_\lambda(\omega))$ are stochastically equivalent weak operators, it follows that, if the first of them is strong, then so is the second. Note, furthermore, that for $x \in \mathbf{D}$

$$|A(\omega)(E_\mu(\omega) - E_\lambda(\omega))\, x| \leqslant [\,|\lambda| \vee |\mu|\,]\,|x|$$

by (2.6), (2.2). Since both sides of (2.6), (2.7) are continuous in x for $-\infty < \lambda < \mu < +\infty$ and $A(\omega)(E_\mu(\omega)) - E_\lambda(\omega)) \in \mathbf{L}(\Omega, \mathbf{X})$, it follows that (2.6) and (2.7) are valid for all $x \in \mathbf{X}$ with probability 1. Since

$$\int_\lambda^\mu \sigma^2 d(E_\sigma(\omega)\, x, x) = |A(\omega)(E_\mu(\omega) - E_\lambda(\omega))\, x|^2 =$$

$$= |(E_\mu(\omega) - E_\lambda(\omega))\, A(\omega)\, x|^2 \leqslant |A(\omega)\, x|^2,$$

we have

$$\mathbf{P}\left\{\int_{-\infty}^{\infty} \sigma^2 d(E_\sigma(\omega)\, x, x) < \infty\right\} = 1.$$

We now show that (2.7) is also valid when $-\infty \leqslant \lambda < \mu \leqslant \leqslant +\infty$. It will suffice to consider the cases: (a) $\lambda = 0$, $\mu = +\infty$; (b) $\lambda = -\infty$, $\mu = 0$, but since these cases are analogous, we shall deal only with the second. We write

$$|A(\omega)(E_0(\omega) - E_{-n}(\omega))x|^2 = \int_{-n}^{0} \sigma^2 d(E_\sigma(\omega)x, x).$$

By condition (1),

$$|(E_0(\omega) - E_{-n}(\omega))A(\omega)x|^2 =$$
$$= ((E_0(\omega) - E_{-n}(\omega))A(\omega)x,\ (E_0(\omega) - E_{-n}(\omega))A(\omega)x) =$$
$$= |E_0(\omega)A(\omega)x|^2 - |E_{-n}(\omega)A(\omega)x|^2.$$

Therefore, for all n,

$$|E_0(\omega)A(\omega)x|^2 \geqslant \int_{-n}^{0} \sigma^2 d(E_\sigma(\omega)x, x)$$

and, letting $n \to \infty$, we get

$$|E_0(\omega)A(\omega)x|^2 = |A(\omega)E_0(\omega)x|^2 \geqslant \int_{-n}^{0} \sigma^2 d(E_\sigma(\omega)x, x).$$

Let $x_n \in \mathbf{D}$. Using previous results, we obtain

$$|A(\omega)E_0(\omega)x_n|^2 = \int_{-\infty}^{0} \sigma^2 d(E_\sigma(\omega)x_n, x_n). \tag{2.9}$$

When $x_n \to x$, $A(\omega)x_n \to A(\omega)x$ in probability. Therefore,

$$|A(\omega)E_0(\omega)x_n - A(\omega)E_0(\omega)x|^2 =$$
$$= |E_0(\omega)(A(\omega)x_n - A(\omega)x)|^2 \leqslant |A(\omega)x_n - A(\omega)x_0|^2.$$

Thus also $A(\omega)E_0(\omega)x_n \to A(\omega)E_0(\omega)x$ in probability. Using the inequality

$$|d(E_\sigma(\omega)x, x) - d(E_\sigma(\omega)y, y)| \leqslant$$
$$\leqslant |d(E_\sigma(\omega)(x-y), x-y)| + 2|d(E_\sigma(\omega)y, x-y)| \leqslant$$
$$\leqslant d(E_\sigma(\omega)(x-y), x-y) + 2\varepsilon d(E'_\sigma(\omega)y, y) +$$
$$+ \frac{2}{\varepsilon} d(E_\sigma(\omega)(x-y), x-y),$$

we obtain

$$\left| \int_{-\infty}^{0} \sigma^2 d(E_\sigma(\omega)x_n, x_n) - \int_{-\infty}^{0} \sigma^2 d(E_\sigma(\omega)x, x) \right| \leqslant$$

$$\leqslant \left(1+\frac{2}{\varepsilon}\right)\int_{-\infty}^{0}\sigma^2 d\,(E_\sigma(\omega)(x_n-x),\ x_n-x)+$$

$$+\,2\varepsilon\int_{-\infty}^{0}\sigma^2 d\,(E_\sigma(\omega)\,x,\ x)\leqslant\left(1+\frac{2}{\varepsilon}\right)|\,A(\omega)\,x_n-A(\omega)\,x\,|\ +$$

$$+\,2\varepsilon\,|\,A(\omega)\,x\,|.$$

As $\varepsilon > 0$ is arbitrary, the right-hand side converges to zero in probability. Hence the right-hand side of (2.9) converges in probability to $\int_{-\infty}^{0}\sigma^2 d\,(E_\sigma(\omega)\,x,\ x)$, while the left-hand side converges in probability to $|\,A(\omega)\times E_0(\omega)\,x\,|$. Formula (2.6) is thus proven for all λ and $\mu\in[-\infty,\ \infty]$.

The extension of formula (2.7) to $\lambda=-\infty$ or $\mu=+\infty$. is analogous. For $x\in\mathbf{D}$

$$(A(\omega)\,E_0(\omega)\,x,\ y)=\int_{-\infty}^{0}\sigma d\,(E_\sigma(\omega)\,x,\ x). \qquad (2.10)$$

Let $x_n\in\mathbf{D},\ x_n\to x$. Then

$$|\,(A(\omega)\,E_0(\omega)\,x_n,\ x_n)-(A(\omega)\,E_0(\omega)\,x,\ x)\,|\leqslant$$
$$\leqslant|\,E_0(\omega)\,A(\omega)\,x_n-E_0(\omega)\,A(\omega)\,x\,|\,|\,x_n\,|+$$
$$+\,|\,E_0(\omega)\,A(\omega)\,x\,|\,|\,x-x_n\,|\to 0,$$

$$\left|\int_{-\infty}^{0}\sigma d\,(E_\sigma(\omega)\,x,\ x)-\int_{-\infty}^{0}\sigma d\,(E_\sigma(\omega)\,x_n,\ x_n)\right|\leqslant$$

$$\leqslant\int_{-\infty}^{0}\frac{1+\sigma^2}{2}\,|\,d\,(E_\sigma(\omega)\,x,\ x)-d\,(E_\sigma(\omega)\,x_n,\ x_n)\,|\to 0$$

in probability as $n\to\infty$. Substituting x_n for x in (2.10) and going to the limit, we see that (2.10) is valid for all $x\in\mathbf{X}$. We have thus proved the following theorem.

Theorem. If $A(\omega)$ is a symmetric operator in $\mathbf{L}_s(\Omega,\ \mathbf{X})$, then there exists a family of projections $\{E_\lambda(\omega),\ \lambda\in\mathbf{R}\}$ in $\mathbf{L}(\Omega,\ \mathbf{X})$, satisfying conditions (1) and (2), for which formulas (2.5)-(2.7) are valid for all $x\in\mathbf{X}$ and $-\infty\leqslant\lambda<\mu\leqslant+\infty$.

Remark. Since

$$|\,d\,(E_\sigma(\omega)\,x,\ y)\,|\leqslant\frac{1}{2}\,[d\,(E_\sigma(\omega)\,x,\ x)+d\,(E_\sigma(\omega)\,y,\ y)],$$

the following integral is defined:

$$\int_{\lambda}^{\mu} \sigma d(E_{\sigma}(\omega)x, y) = (A(\omega)(E_{\mu}(\omega) - E_{\lambda}(\omega))x, y).$$

Remark 2. The integral

$$\int_{-\infty}^{\infty} \sigma dE_{\sigma}(\omega)x$$

is defined for all $x \in \mathbf{X}$ and almost all ω, and in fact

$$\int_{-\infty}^{\infty} \sigma dE_{\sigma}(\omega)x = A(\omega)x. \tag{2.11}$$

5. Equations with Symmetric Random Operators

5.1. Evolution Equations

Consider an equation of the form

$$\frac{dx_t(\omega)}{dt} = A(\omega)x_t(\omega), \quad t \geqslant 0, \quad x_0 \in \mathbf{X}, \tag{2.12}$$

where $x_t(\omega)$ is an unknown function with values in $\mathbf{X}(\Omega)$, $A(\omega)$ is a symmetric operator in $\mathbf{L}_s(\Omega, \mathbf{X})$. Equation (2.12) should be understood in the weak sense, i.e., we wish to find a function $x_t(\omega)$, such that for all

$$\frac{d}{dt}(x_t(\omega), x) = (A(\omega)x, x_t(\omega)), \quad t > 0;$$

$$(x_0(\omega), x) = (x_0, x). \tag{2.13}$$

We shall utilize the spectral family $E_{\lambda}(\omega)$ whose existence was proved in § 4. Suppose that for $t > 0$ the following expression is finite:

$$\gamma_t(\omega, x) = \int e^{\lambda t} d(E_{\lambda}(\omega)x, x). \tag{2.14}$$

If, moreover,

$$\mathbf{P}\left\{\int |\lambda| e^{\lambda t} d(E_\lambda(\omega)x, x) < \infty\right\} = 1, \tag{2.15}$$

then $\gamma_t(\omega, x)$ is differentiable with respect to t and

$$\frac{d}{dt}\gamma_t(\omega, x) = \int \lambda e^{\lambda t} d(E_\lambda(\omega)x, x).$$

Define $y_t(\omega, x)$ by the equality

$$(y_t(\omega, x), z) = \int e^{\lambda t} d(E_\lambda(\omega)x, z). \tag{2.16}$$

then

$$|y_t(\omega, x)|^2 = \int e^{2\lambda t} d(E_\lambda(\omega)x, x)$$

and by (2.14) $y_t(\omega)$ is an $\mathbf{X}$-valued r.v. with probability 1. Furthermore,

$$(y_t(\omega, x), A(\omega)x) = \int \lambda e^{\lambda t} d(E_\lambda(\omega)x, x);$$

$$(y_t(\omega, x), x) = \int e^{\lambda t} d(E_\lambda(\omega)x, x).$$

Consequently,

$$\frac{d}{dt}(y_t(\omega, x), x) = (y_t(\omega, x), A(\omega)x).$$

But then, for all $x_0 \in \mathbf{X}$, $x \in \mathbf{X}$

$$\frac{d}{dt}(y_t(\omega, x_0), x) = (y_t(\omega, x_0), A(\omega)x).$$

Thus, if condition (2.15) holds for $x = x_0$, the solution of equation (2.12) is given by the formula

$$x_t(\omega) = \int e^{\lambda t} dE_\lambda(\omega)x_0. \tag{2.17}$$

Note that a sufficient condition for (2.15) to hold is that for some $\gamma(\omega)$

$$(A(\omega)x, x) \leqslant \gamma(\omega)(x, x), \tag{2.18}$$

since in that case $\int\limits_{\gamma(\omega)}^{\infty} d(E_\lambda(\omega)x, x) = 0$ for all $x \in \mathbf{X}$. Thus

$$\int\limits_{-\infty}^{\infty} |\lambda| e^{\lambda t} d(E_\lambda(\omega)x, x) = \int\limits_{-\infty}^{\gamma(\omega)} |\lambda| e^{\lambda t} d(E_\lambda(\omega)x, x),$$

and the function $|\lambda| e^{\lambda t}$ $(t > 0)$ is bounded on $(-\infty, \gamma)$. We claim that when condition (2.18) is satisfied, the solution of equation (2.12) satisfying the condition

$$|x_t(\omega)| \leqslant \alpha(\omega) e^{t\gamma^*(\omega)},$$

where $\gamma^*(\omega)$ is $\mathfrak{S}$-measurable, is unique and is therefore given by (2.17). The uniqueness proof need be carried out only for $x_0 = 0$. Let $\lambda > 0$ and $\Gamma_\lambda = \{\omega\colon\ \gamma^*(\omega) + \gamma(\omega) < \lambda\}$. Then, for $\omega \in \Gamma_\lambda$, $x \in \mathbf{X}$,

$$\int\limits_0^\infty e^{-\lambda t} \frac{d}{dt}(x_t(\omega), x)\, dt = \int\limits_0^\infty e^{-\lambda t} (A(\omega)x, x_t(\omega))\, dt. \qquad (2.19)$$

Denote

$$z_\lambda(\omega) = \int\limits_0^\infty e^{-\lambda t} x_t(\omega)\, dt. \qquad (2.20)$$

Then, integrating the left-hand side of (2.19) by parts, we find that

$$\lambda(z_\lambda(\omega), x) = (A(\omega)x, z_\lambda(\omega)). \qquad (2.21)$$

Now, instead of $A(\omega)$ consider the selfadjoint random operator $\bar{A}(\omega)$, constructed in § 4. It follows from (2.21) that for $x \in \mathbf{D}$

$$\lambda(z_\lambda(\omega), x) = (\bar{A}(\omega)x, z_\lambda(\omega)).$$

Thus the operator $\bar{A}(\omega)$ is defined on $z_\lambda(\omega)$ and $\bar{A}(\omega) z_\lambda(\omega) = \lambda z_\lambda(\omega)$. Since

$$(\bar{A}(\omega)x, x) = \int\limits_{-\infty}^{\gamma(\omega)} d(E_\sigma(\omega)x, x) \leqslant \gamma(\omega)(x, x),$$

it follows that

$$(\bar{A}(\omega) z_\lambda(\omega), z_\lambda(\omega)) \leqslant \gamma(\omega) |z_\lambda(\omega)|^2$$

for $\omega \in \Gamma_\lambda$. But $(\bar{A}(\omega) z_\lambda(\omega), z_\lambda(\omega)) = \lambda |z_\lambda(\omega)|^2$ and so, for $\omega \in \Gamma_\lambda$

$$(\gamma(\omega) - \lambda) |z_\lambda(\omega)|^2 \leqslant 0,$$

which is possible only if $|z_\lambda(\omega)| = 0$. Since $\Gamma_\mu \supset \Gamma_\lambda$ when $\lambda < \mu$, it follows that $z_\mu(\omega) = 0$ on Γ_λ for $\mu > \lambda$. Hence it follows, because of (2.20), that $x_t(\omega) = 0$ on Γ_λ, whatever λ. Since $\Omega = \bigcup_\lambda \Gamma_\lambda$, we have $x_t(\omega) = 0$ on Ω. This proves the uniqueness of the solution.

5.2. Schrödinger-type Equations

Let $A(\omega)$ be the same as in the preceding subsection. Consider the equation

$$\frac{dx_t(\omega)}{dt} = i\tilde{A}(\omega) x_t(\omega), \quad t \geqslant 0, \quad x_0(\omega) = x_0, \tag{2.22}$$

where $x_t(\omega)$ takes values in $\tilde{\mathbf{X}}$ and $\tilde{A}(\omega)$ is the natural extension of $A(\omega)$ to $\tilde{\mathbf{X}}$. As before, this equation should be interpreted in the weak sense, that is to say, for all $x \in \tilde{\mathbf{X}}$ we have an equality

$$\frac{d(x_t(\omega), x)}{dt} = i(x_t(\omega), \tilde{A}(\omega) x), \quad t > 0,$$

$$(x_0(\omega), x) = (x_0, x). \tag{2.23}$$

It can be verified that the solution has the form

$$x_t(\omega) = \int e^{it\lambda} dE_\lambda(\omega) x_0,$$

and in this case there is no need to impose additional restrictions of the type of (2.15), since

$$\int \lambda e^{it\lambda} d(E_\lambda(\omega) x, x)$$

is always absolutely convergent together with

$$\int \lambda^2 d(E_\lambda(\omega) x, x).$$

Note, further, that

$$|x_t(\omega)|^2 = \left(\int e^{it\lambda} d\,E_\lambda(\omega)\,x_0,\ \int e^{it\lambda} dE_\lambda(\omega)\,x_0\right) =$$

$$= \int e^{it\lambda}\overline{e^{it\lambda}} d\,(E_\lambda(\omega)\,x_0,\ x_0) = \int d\,(E_\lambda(\omega)\,x_0,\ x_0) = |x_0|^2.$$

Thus, the operator

$$U_t(\omega)\,x_0 = \int e^{it\lambda} dE_\lambda(\omega)\,x_0$$

belongs to $\mathbf{L}(\Omega, \mathbf{X})$ with probability 1 and is unitary.

5.3. Spectral; Moment Functions

The operators $U_t(\omega)$ constructed in the preceding subsection can be utilized to compute the spectral moment functions

$$\mathfrak{M}_{\lambda_1,\ldots,\lambda_k}(C_1, \ldots, C_k) =$$

$$= \mathbf{M}\,\mathrm{sp}\,(E_{\lambda_1}(\omega)\,C_1) \ldots \mathrm{sp}\,(E_{\lambda_k}(\omega)\,C_k), \tag{2.24}$$

where $C_1, \ldots, C_k \in \mathbf{L}_0(\mathbf{X})$. In particular, the first moment function is defined by the operator

$$\mathcal{E}(\lambda) = \mathbf{M}E_\lambda(\omega).$$

The existence of the expectation in this formula (and in (2.24)) follows from the boundedness of $E_\lambda(\omega)$. Once the operator $\mathcal{E}(\lambda)$ is known, one can evaluate certain moment functions of the solution of equation (2.12), e.g.,

$$\mathbf{M}x_t(\omega) = \int e^{\lambda t} d\mathcal{E}(\lambda)\,x_0;$$

$$\mathbf{M}\,|x_t(\omega)|^2 = \int e^{2\lambda t} d\,(\mathcal{E}(\lambda)\,x_0,\ x_0);$$

$$\mathbf{M}\,(x_t(\omega),\ x_s(\omega)) = \int e^{(t+s)\lambda} d\,(\mathcal{E}(\lambda)\,x_0,\ x_0).$$

Since $\mathcal{E}(\lambda)$ is a symmetric operator, evaluation of this function requires only a knowledge of $(\mathcal{E}(\lambda)\,x, x)$ for any x. This is a nondecreasing function. Its Fourier-Stieltjes transform is

$$\int e^{i\lambda t} d\,(\mathcal{E}(\lambda)\,x,\ x) = \mathbf{M}\,(U_t(\omega)\,x,\ x). \tag{2.25}$$

Thus, in order to evaluate $\mathcal{E}(\lambda)$ it will suffice to evaluate the expression on the left of (2.25). If the operator $A(\omega)$ in equation (2.22) is bounded by a nonrandom constant, then

$$\mathbf{M}(U_t(\omega)x, x) = \sum_{n=0}^{\infty} \frac{(it)^n}{n!} \mathbf{M}(A^n(\omega)x, x).$$

In the general case, one can use the following proposition.

Lemma. Let $A(\omega) \in \mathbf{L}_s(\Omega, \mathbf{X})$ be symmetric, $A_m(\omega) \in \mathbf{L}(\Omega, \mathbf{X})$, $\|A_m(\omega)\| < m$, $A_m(\omega)$ symmetric and $A_m(\omega)x \to A(\omega)x$ in probability $\forall x \in \mathbf{X}$. Let $U_t^{(m)}(\omega)x_0$ denote the solution of equation (2.22) with $A_m(\omega)$ in place of $A(\omega)$. Then $U_t^{(m)}(\omega)x_0 \to U_t(\omega)x_0$ in probability.

Proof. Note that

$$|U_t^{(m)}(\omega)x_0 - U_{t+h}^{(m)}(\omega)x_0|^2 = |U_h^{(m)}(\omega)x_0 - x_0|^2 =$$
$$= 2|x_0|^2 - 2\operatorname{Re}(U_h^{(m)}(\omega)x_0, x_0) =$$
$$= 2\int (1 - \cos h\lambda)\, d(E_\lambda^{(m)}(\omega)x_0, x_0) \leqslant$$
$$\leqslant h^2 \int \lambda^2 d(E_\lambda^{(m)}(\omega)x_0, x_0) = h^2 |A_m(\omega)x_0|^2,$$

where $E_\lambda^{(m)}(\omega)$ is a spectral family for $A_m(\omega)$. Thus the functions $U_t^{(m)}(\omega)x_0$ are continuous in t, uniformly with respect to m. Next, for $\gamma > 0$

$$\int_0^\infty e^{-\gamma t} U_t^{(m)}(\omega)x_0 dt = \int_0^\infty e^{-\gamma t + itA_m(\omega)} x_0 dt = (\gamma I - iA_m(\omega))^{-1} x_0.$$

The fact that $U_t^{(m)}(\omega)$ is unitary implies that

$$\|(\gamma I - iA_m(\omega))^{-1}\| \leqslant \frac{1}{\gamma}. \tag{2.26}$$

Similarly,

$$\int_0^\infty e^{-\gamma t} U_t(\omega)\, dt = (\gamma I - i\bar{A}(\omega))^{-1}$$

(here $\bar{A}(\omega)$ is the selfadjoint extension of $A(\omega)$) and $\|(\gamma I - iA(\omega))^{-1}\| \leqslant \frac{1}{\gamma}$. Therefore, in order to prove that

$$(\gamma I - iA_m(\omega))^{-1} x \to (\gamma I - i\bar{A}(\omega))^{-1} x$$

in probability for all $x \in \mathbf{X}$ it will suffice to prove this for some dense set. The set of all y expressible as

$$y = (\gamma I - i\bar{A}(\omega))\,x,$$

where $x \in \mathbf{X}$, is dense in $\mathbf{X}$ (this was proved in § 4 for $\gamma = 1$; for arbitrary $\gamma > 0$, consider the operator $\frac{1}{\gamma}\bar{A}(\omega)$. We have

$$(\gamma I - iA_m(\omega))^{-1} y - (\gamma I - i\bar{A}(\omega))^{-1} y =$$
$$= i(\gamma I - iA_m(\omega))^{-1}[\bar{A}(\omega)x - A_m(\omega)x].$$

This expression tends to zero in view of the identity of $A(\omega)x$ and $A(\omega)x$ for $x \in \mathbf{D}$, the assumption of the lemma and inequality (2.26). Thus, the Laplace transforms of the uniformly continuous functions $U_t^{(m)}(\omega)x$ are convergent, whence it follows that the functions themselves are also convergent.■

<u>Remark</u>. If $A_m(\omega)$ satisfy the assumptions of the lemma, the higher-order spectral moment functions may be evaluated with the help of their Fourier transform:

$$\int \ldots \int d_{\lambda_1}, \ldots, d_{\lambda_k} \mathfrak{M}_{\lambda_1,\ldots,\lambda_k}(C_1, \ldots, C_k)\, e^{i\lambda_1 t_1 + \cdots + i\lambda_k t_k} =$$
$$= \mathbf{M}(\mathrm{sp}\, U_{t_1}(\omega) C_1) \ldots (\mathrm{sp}\, U_{t_k}(\omega) C_k), \qquad (2.27)$$

where the expression on the right is defined as the limit

$$\lim_{m\to\infty} \mathbf{M}(\mathrm{sp}\, U_{t_1}^{(m)}(\omega) C_1) \ldots (\mathrm{sp}\, U_{t_k}^{(m)}(\omega) C_k),$$

and $U_t^{(m)}(\omega)$ may be evaluated as

$$U_t^{(m)}(\omega) = \sum_{n=0}^{\infty} \frac{(it)^n}{n!} (A_m(\omega))^n.$$

5.4. Equation of Fredholm Type

Consider the equation

$$x(\omega) = x + \alpha A(\omega)\, x(\omega), \qquad (2.28)$$

where $A(\omega)$ is a symmetric random operator in $\mathbf{L}_s(\Omega, \mathbf{X})$, $x \in \mathbf{X}$, $x(\omega)$ is an unknown r.v. with values in $\mathbf{X}(\Omega)$, α is some real number. This equation may also be considered for complex α, in which case $x(\omega)$ takes values in $\tilde{\mathbf{X}}(\Omega)$ and $A(\omega)$ should be replaced by its natural extension from $\mathbf{X}$ to $\tilde{\mathbf{X}}$. Equation (2.28) should be interpreted in the following sense:

$$(x(\omega),\ y) = (x,\ y) + \alpha\,(x(\omega),\ A(\omega)\,y),\quad \forall\, y \in \mathbf{X}. \tag{2.29}$$

Using the spectral representation, we can reduce the equation to the following form:

$$(x(\omega),\ y) = (x,\ y) + \alpha \int \lambda\,(dE_\lambda(\omega)\,y,\ x(\omega));$$

$$(x,\ y) = \int (\alpha\lambda - 1)\, d\,(E_\lambda(\omega)\,y,\ x(\omega)) =$$

$$= \int (\alpha\lambda - 1)\, d\,(E_\lambda(\omega)\,x(\omega),\ y),$$

whence

$$x = \int (\alpha\lambda - 1)\, dE_\lambda(\omega)\,x(\omega),$$

$$x(\omega) = \int \frac{1}{\alpha\lambda - 1}\, dE_\lambda(\omega)\,x. \tag{2.30}$$

If the integral on the right exists, it is readily checked that $x(\omega)$ as defined by formula (2.30) is indeed a solution of (2.29). When is the solution of equation (2.29) unique? Consider the equation at $x = 0$. If $x(\omega)$ is a solution, then

$$\frac{1}{\alpha}(x(\omega),\ y) = (x(\omega),\ A(\omega)\,y)$$

and so $\frac{1}{\alpha}$ is an eigenvalue of the selfadjoint extension $\tilde{A}(\omega)$ of $A(\omega)$. If α is complex, $\operatorname{Im}\alpha \neq 0$, this is impossible. Let α be real. Let $\lambda_i(\omega)$ denote the points of the discrete spectrum of $\tilde{A}(\omega)$,

$$\Gamma_\alpha = \left\{\omega\colon \text{ for some } i\ \ \alpha = \frac{1}{\lambda_i(\omega)}\right\}.$$

There can be at most countably many points α, such that $\mathbf{P}\{\Gamma_\alpha\} > 0$. Indeed, let α be such that $(\mathcal{E}(\lambda)\,x,\ x)$, where $\mathcal{E}(\lambda) = \mathbf{M}E_\lambda(\omega)$, is continuous in λ at the point $1/\alpha$, for all x in some countable dense subset of $\mathbf{X}$. Then also $\mathbf{M}\{(E_{\lambda_1}(\omega)\,x,\ x) - (E_{\lambda_2}(\omega)\,x,\ x)\} = (\mathcal{E}(\lambda_1)\,x, x) - (\mathcal{E}(\lambda_2)\,x,\ x) \to 0$ as $\lambda_1 \downarrow 1/\alpha$, $\lambda_2 \uparrow 1/\alpha$. Thus, it is also true that $(E_{\lambda_1}(\omega)\,x, x) - (E_{\lambda_2}(\omega)\,x, x) \to 0$ as $\lambda_1 \downarrow 1/\alpha$, $\lambda_2 \uparrow 1/\alpha$, since $(E_\lambda(\omega)\,x,\ x)$ is a nondecreasing function. If $\omega \in \Gamma_\alpha$, then there exists $z(\omega)$ such that

$$(E_{\lambda_1}(\omega)z(\omega),\ z(\omega)) - (E_{\lambda_2}(\omega)z(\omega),\ z(\omega)) \downarrow (z(\omega),\ z(\omega)) \quad (2.31)$$

as $\lambda_1 \downarrow 1/\alpha$, $\lambda_2 \uparrow 1/\alpha$. If x is chosen in such a way that $\mathbf{P}\{|x - z(\omega)| < \varepsilon\} > 0$, and x is such that $(E_\lambda(\omega)x, x)$ is continuous at $1/\alpha$, $|x| = 1$, then, as $\mathbf{P}\{|(E_\lambda(\omega)x, x) - (E_\lambda(\omega) \times z(\omega), z(\omega))| \leqslant 2\varepsilon\} > 0$, the discontinuity of $E_\lambda(\omega)z(\omega)$ at the point $1/\alpha$ is at most 4ε; this contradicts (2.31) with $\varepsilon < 1/4$. Note that the points of continuity of $(\mathscr{E}(\lambda)x, x)$ for all x in some countable dense set coincide with the points of weak continuity (with respect to λ) of the operator-valued function $\mathscr{E}(\lambda)$. Thus, equation (2.28) has a unique solution for all α, such that $1/\alpha$ is a weak continuity point of $\mathscr{E}(\lambda)$ (i.e., $(\mathscr{E}(\lambda)x, y)$ is continuous for all x, y). This solution - if it exists - has the form (2.30). The right-hand side of (2.30) is defined as an $\mathbf{X}$ -valued r.v. if and only if

$$\int \left(\frac{1}{\alpha\lambda - 1}\right)^2 d(E_\lambda(\omega)x,\ x) < \infty. \quad (2.32)$$

Remark. Inequality (2.32) will hold if

$$\int \left(\frac{1}{\alpha\lambda - 1}\right)^2 d(\mathscr{E}(\lambda)x,\ x) < \infty. \quad (2.33)$$

One condition for (2.33) to hold is as follows. Consider the spectral density

$$\rho_\lambda(x) = \frac{d}{d\lambda}(\mathscr{E}(\lambda)x,\ x).$$

Assume that $\rho_\lambda(x)$ is twice continuously differentiable with respect to λ. Then inequality (2.33) is true if and only if $\rho_{1/\alpha}(x) = 0$.

6. Equations with Semi-bounded Random Operators

6.1. Nonnegative Closed Random Operators

The results obtained in § 5 for evolution equations and equations of Fredholm type with symmetric random operators may be extended to semi-bounded random operator $A(\omega) \in \mathbf{L}_\omega(\Omega, \mathbf{X})$ is said to be semi-bounded if there exists an $\mathfrak{S}$ -measurable r.v. $\gamma(\omega)$, such that, with probability 1.

$$\forall x \qquad (A(\omega)x, x) \leqslant \gamma(\omega)(x, x). \tag{2.34}$$

Replacing $A(\omega)$ by the operator $\gamma(\omega)I - A(\omega)$, one can always reduce the study of semi-bounded operators to that of non-negative operators.

In this section we shall consider semi-bounded or non-negative operators such that, relative to some basis $\{e_k\}$ the following series are convergent:

$$\forall i \ \sum_{k=1}^{\infty}(A(\omega)e_i, e_k)^2 < \infty; \ \sum_{k=1}^{\infty}(A(\omega)e_k, e_i)^2 < \infty. \tag{2.35}$$

Under this condition one can define random variables

$$A(\omega)e_i = \sum_{k=1}^{\infty}(A(\omega)e_i, e_k)e_k, \quad A^*(\omega)e_i = \sum_{k=1}^{\infty}(A(\omega)e_k, e_i)e_k. \tag{2.36}$$

Moreover, let $\sigma_i > 0$ be such that

$$\sum_{i,k}(A(\omega)e_i, A(\omega)e_k)^2\sigma_i^2\sigma_k^2 < \infty;$$

$$\sum_{i,k}(A^*(\omega)e_i, A^*(\omega)e_k)^2\sigma_i^2\sigma_k^2 < \infty.$$

If $B \in \mathbf{L}(\mathbf{X})$ and $Be_k = \sigma_k e_k$, then $C(\omega) = BA(\omega)B$ and $C^*(\omega) = BA^*(\omega)B$ are stochastically equivalent operators in $\mathbf{L}(\Omega, \mathbf{X})$ (the proof is the same as in subsection 1, § 4). These operators are adjoint to one another and their ranges for all ω are subsets of $\mathbf{D}$ - the range of B. By $C(\omega)$ and $C^*(\omega)$ we mean operators in $\mathbf{L}(\Omega, \mathbf{X})$. Put

$$\tilde{A}(\omega) = B^{-1}C(\omega)B^{-1}, \quad \tilde{A}^*(\omega) = B^{-1}C^*(\omega)B^{-1}.$$

For any ω these operators are defined on $\mathbf{D}$, and then, for $x \in \mathbf{D}$, $y \in \mathbf{X}$,

$$(\tilde{A}(\omega)x, y) = (A(\omega)x, y), \quad (\tilde{A}^*(\omega)x, y) = (A^*(\omega)x, y) =$$

$$= (x, A(\omega)y)$$

with probability 1. The operators $\tilde{A}(\omega)$ and $\tilde{A}^*(\omega)$ have closures from the set $\mathbf{D}$. This follows from the following proposition.

Lemma 1. Let C be a bounded operator, C^* its adjoint,

where the range of C and C^* is a subset of $\mathbf{D}$; let B be a symmetric operator with range $\mathbf{D}$. Then the operator $B^{-1}CB^{-1}$, which is defined on $\mathbf{D}$, has a closure.

Proof. For $x, y \in \mathbf{D}$, we have

$$(B^{-1}CB^{-1}x, y) = (x, B^{-1}C^*B^{-1}y),$$

and so $B^{-1}CB^{-1}$ coincides with the closed operator $[B^{-1}C^*B^{-1}]^*$ (the adjoint to $B^{-1}C^*B^{-1}$); hence it has a closure.

Denote the closures of $\tilde{A}(\omega)$ and $\tilde{A}^*(\omega)$, considered as operators with domain $\mathbf{D}$, by $\hat{A}(\omega)$ and $\hat{A}^*(\omega)$, respectively. These are identical to $\tilde{A}(\omega)$ and $\tilde{A}^*(\omega)$ on $\mathbf{D}$. Then the domains of $\hat{A}(\omega)$ and $\hat{A}^*(\omega)$ depend on ω. One must bear in mind that the operators $\hat{A}(\omega)$ and $\hat{A}^*(\omega)$ are not necessarily adjoint for all ω.

Assume that $A(\omega)$ is a nonnegative operator. Then $\tilde{A}(\omega)$ is also nonnegative on $\mathbf{D}$; since

$$(\tilde{A}(\omega)x, x) = (\tilde{A}^*(\omega)x, x),$$

the same is true of the operator $\tilde{A}^*(\omega)$. Hence the operators $\hat{A}(\omega)$ and $\hat{A}^*(\omega)$, as closures of nonnegative operators, are nonnegative in their domain of definition. We shall now construct the resolvent of $\hat{A}(\omega)$; first we consider nonrandom operators.

6.2. Resolvent of a Nonnegative Operator

Let A be a nonrandom closed operator with domain $\mathbf{D}$, dense in $\mathbf{X}$, such that $(Ax, x) \geqslant 0$ for $x \in \mathbf{D}$. Let $\mathbf{L}$ denote the range of the operator $A + I$; $\mathbf{L}$ is a linear subspace. If $x_n \in \mathbf{L}$, then $x_n = y_n + Ay_n$. Moreover, we have

$$|x_n|^2 = (y_n + Ay_n, y_n + Ay_n) =$$

$$= |y_n|^2 + 2(Ay_n, y_n) + |Ay_n|^2 \geqslant |y_n|^2 + |Ay_n|^2.$$

The fact that x_n converges to a limit x_0 implies that $y_n \to y_0$, $Ay_n \to z_0$, and since the operator is closed this yields $z_0 = Ay_0$ and $x_0 = y_0 + Ay_0$. Thus $\mathbf{L}$ is closed. We shall say that the operator A is maximal if $\mathbf{L} = \mathbf{X}$. We claim that in this case, for all $\lambda > 0$, the range of the operator $\lambda I + A$ coincides with $\mathbf{X}$. Since

$$((\lambda I + A)x, (\lambda I + A)x) = \lambda^2 |x|^2 + 2\lambda (Ax, x) + |Ax|^2 \geqslant$$
$$\geqslant \lambda^2 |x|^2,$$

it follows that for all $\lambda > 0$ the operator $R_\lambda = (\lambda I + A)^{-1}$ is defined; moreover, the norm of this operator satisfies the inequality $\| R_\lambda \| \leqslant \frac{1}{\lambda}$. We shall call this operator the resolvent of A. By assumption, the operator R_1 is defined on $\mathbf{X}$. Let $|\lambda - 1| < 1$; then the equation

$$(\lambda I + A)x = y \tag{2.37}$$

is solvable for all $y \in \mathbf{X}$. Indeed, letting $x = R_1 z$, we obtain

$$(\lambda - 1) R_1 z + z = y;$$
$$z = \sum_{n=0}^{\infty} (1 - \lambda)^n R_1^n y$$

(the convergence of the series follows from the inequalities $\| (1 - \lambda) R_1 \| \leqslant |1 - \lambda| < 1$). The above formula yields an expression for the resolvent when $|1 - \lambda| < 1$

$$R_\lambda y = \sum_{n=0}^{\infty} (1 - \lambda)^n R_1^{n+1} y.$$

One shows in similar fashion that when R_{λ_1} is defined on $\mathbf{X}$ then, if $|\lambda - \lambda_1| < \lambda_1$,

$$R_\lambda y = \sum_{n=0}^{\infty} (\lambda_1 - \lambda)^n R_{\lambda_1}^{n+1} y \tag{2.38}$$

(the convergence of the series follows from the inequalities $\left. \| (\lambda_1 - \lambda) R_{\lambda_1} \| \leqslant \frac{|\lambda_1 - \lambda|}{\lambda_1} < 1 \right)$. Formula (2.38) enables us to define the resolvent for all $\lambda > 0$ (at the same time automatically establishing that equation (2.37) is solvable for all $\lambda > 0$). Moreover, using formula (2.38) one can show that the resolvent exists for all complex λ with $\lambda > 0$.

For all positive λ the operator R_λ is also nonnegative. Indeed, let $x = \lambda y + Ay$ (by what we have proved, all elements $x \in \mathbf{X}$ can be expressed in this form). Then

$$(R_\lambda x, x) = (R_\lambda(\lambda y + Ay), \lambda y + Ay) = (y, \lambda y + Ay) =$$
$$= \lambda |y|^2 + (y, Ay) \geqslant 0.$$

Remark. We have defined the resolvent in this section as $(\lambda I + A)^{-1}$, whereas in the generally accepted terminology this operator is the resolvent of $-A$. However, our definition is more natural for nonnegative operators. To avoid confusion, we shall always exhibit the formula defining the resolvent.

We shall now show that any closed nonnegative operator may be extended to a maximal nonnegative operator. Suppose that $\mathbf{L}$ is not all of $\mathbf{X}$. Let $\mathbf{N}$ denote the orthogonal complement to $\mathbf{L}$. Then for $z \in \mathbf{N}$

$$0 = (Ax + x, z), \ (x, z) = -(Ax, z) \qquad \text{for all} \quad x \in \mathbf{D}.$$

Thus A^*z, is defined, $(x, z + A^*z) = 0 \ \forall x \in \mathbf{D}$, and so $A^*z = -z$. Therefore, if $z \in \mathbf{N} \cap \mathbf{D}$ then

$$0 \leqslant (Az, z) = (z, A^*z) = -|z|^2, \quad z = 0.$$

Define the operator $\tilde{A}$ on $\mathbf{N} + \mathbf{D}$ as follows. If $x = y + z$, $y \in \mathbf{D}$, $z \in \mathbf{N}$, then

$$\bar{A}x = Ay + z.$$

We claim that $\bar{A}$ is closed. Let $x_n = y_n + z_n$, $x_n \to x_0$, $\bar{A}x_n \to x_0$. This implies the convergence of the sequence

$$x_n + \bar{A}x_n = y_n + Ay_n + 2z_n.$$

Since $y_n + Ay_n \in \mathbf{L}$, $2z_n \in \mathbf{N}$, there exist limits $\lim z_n = z_0 \in \mathbf{N}$, $\lim y_n = x_0 - z_0$. In addition, the limit $Ay_n = \bar{A}x_n - z_n$ exists. Therefore, $y_0 = x_0 - z_0 \in \mathbf{D}$, $\lim Ay_n = Ay_0$ and $\lim \bar{A}x_n = y_0 + z_0 = \bar{A}x_0$. Finally, we show that $\bar{A}$ is a nonnegative operator. For $x = y + z$, $y \in \mathbf{D}$, $z \in \mathbf{N}$,

$$(\bar{A}x, x) = (Ay + z, y + z) =$$
$$= (Ay, y) + (z, z) + (z, y) + (Ay, z) = (Ay, y) + (z, z) \geqslant 0,$$

since $(Ay + y, z) = 0 \quad (Ay + y \in \mathbf{L}, z \in \mathbf{N})$.

Note that the maximal extension of an operator need not be unique.

6.3. Resolvent of a Nonnegative Random Operator

Let $A(\omega)$ be a nonnegative operator satisfying the conditions of subsection 1, § 6, and $\hat{A}(\omega)$ is a closed operator identical with $A(\omega)$ on $\mathbf{D}$. It is convenient to take $\mathbf{D}$ to be the linear span of the vectors $\{e_k, k = 1, 2, \ldots\}$. Let $\bar{A}(\omega)$ denote a maximal operator derived from $\hat{A}(\omega)$ as described in the previous subsection. Let $\bar{R}_\lambda(\omega, A) = (\lambda I + \bar{A}(\omega))^{-1}$. For any ω, this operator belongs to $\mathbf{L}(\mathbf{X})$, $\|\bar{R}_\lambda(\omega, A)\| \leqslant \frac{1}{\lambda}$. We shall prove that $\bar{R}_\lambda(\omega, A) \in \mathbf{L}(\Omega, \mathbf{X})$, i.e., for all $x, y \in \mathbf{X}$ the function $(\bar{R}_\lambda(\omega, A)x, y)$ is $\mathfrak{S}$-measurable. Using formula (2.38), one readily sees that it will suffice to prove $\mathfrak{S}$-measurability for $(\bar{R}_1(\omega, A)x, y)$. Let $Q(\omega)$ denote projection onto subspace $\mathbf{N}_\omega$ orthogonal to the set $\{z : z = x + \hat{A}(\omega)x\}$ (ω fixed). As shown in subsection 2, § 5, $Q(\omega) \in \mathbf{L}(\Omega, \mathbf{X})$. Note that by construction $\bar{R}_1(\omega, A)y$ is the unique solution of the equation $x + \bar{A}(\omega)x = y$; it is easy to see that if $y \in \mathbf{N}_\omega$ the solution is $x = \frac{1}{2}y$. Thus, if $y \in \mathbf{N}_\omega$ then also $\bar{R}_1(\omega, A)y \in \mathbf{N}_\omega$ with $\bar{R}_1(\omega, A)y = \frac{1}{2}y$. Therefore,

$$\bar{R}_1(\omega, A)Q(\omega)x = \frac{1}{2}Q(\omega)x.$$

Now, $x - Q(\omega)x$ lies in the closure of the set

$$\mathbf{L}_\omega^0 \doteq \{z:\ z = x + \bar{A}(\omega)x,\ x \in \mathbf{D}\}.$$

Since the linear combinations of the vectors $z_k(\omega) = e_k + \tilde{A}(\omega)e_k$ are dense in $\mathbf{L}_\omega^0$, it follows that $x_2(\omega) = \bar{R}_1(\omega, A)(I - Q(\omega))x$, which is a solution of the equation

$$\dot{x}_2(\omega) + \hat{A}(\omega)x_2(\omega) = (I - Q(\omega))x,$$

may be constructed as follows. We first orthogonalize the sequence $z_k(\omega)$; obtaining, say, vectors $y_1(\omega), y_2(\omega), \ldots$. Let

$$\xi_k(\omega) = ((I - Q(\omega))x,\ y_k(\omega)).$$

Since $\lim\limits_{n\to\infty}\left| x_2(\omega) + \hat{A}(\omega)\, x_2(\omega) - \sum\limits_{k=1}^{n} \xi_k(\omega)\, y_k(\omega)\right| = 0$, we can find elements $x^{(n)}(\omega) \in \mathbf{D}$, such that

$$\sum_{k=1}^{n} \xi_k(\omega)\, y_k(\omega) = x^{(n)}(\omega) + \breve{A}(\omega)\, x^{(n)}(\omega),$$

and then

$$\overline{\lim_{n\to\infty}}\, | x_2(\omega) - x^{(n)}(\omega) | \leqslant \lim_{n\to\infty} | x_2(\omega) + \hat{A}(\omega)\, x_2(\omega) -$$

$$- x^{(n)}(\omega) - \hat{A}(\omega)\, x^{(n)}(\omega) | = 0.$$

If $y_k(\omega) = \sum\limits_{j=1}^{k} \alpha_{kj}(\omega)\, z_j(\omega)$, then

$$\sum_{k=1}^{n} \xi_k(\omega)\, y_k(\omega) = \sum_{k=1}^{n} \xi_k(\omega) \sum_{j=1}^{k} \alpha_{kj}(\omega)\, z_j(\omega),$$

so that

$$x^{(n)}(\omega) = \sum_{k=1}^{n} \xi_k(\omega) \sum_{j=1}^{k} \alpha_{kj}(\omega)\, e_j.$$

Since $\xi_k(\omega)$ and $\alpha_{kj}(\omega)$ are $\mathfrak{S}$-measurable r.v.'s, it follows that $x^{(n)}(\omega) \in \mathbf{X}(\Omega)$, and so also $x_2(\omega) \in \mathbf{X}(\Omega)$. We have thereby proved that

$$\bar{R}_1(\omega, A)\, x = x_2(\omega) + \frac{1}{2} Q(\omega)\, x \in \mathbf{X}(\Omega)$$

or $\bar{R}_1(\omega, A) \in \mathbf{L}(\Omega, \mathbf{X})$.

By definition of the resolvent $\bar{R}_\lambda(\omega, A)$, the following relationships hold:

$$\lambda \bar{R}_\lambda(\omega, A)\, x + \bar{R}_\lambda(\omega, A)\, \bar{A}(\omega)\, x = x, \quad x \in \mathbf{D} + \mathbf{N}_\omega; \qquad (2.39)$$

$$\lambda \bar{R}_\lambda(\omega, A)\, x + \bar{A}(\omega)\, \bar{R}_1(\omega, A)\, x = x, \quad x \in \mathbf{X}. \qquad (2.40)$$

It follows from (2.39) and from the equality $A(\omega)\, x = \bar{A}(\omega)\, x$ for $x \in \mathbf{D}$ that

$$\lambda \bar{R}_\lambda(\omega, A)\, x + \bar{R}_\lambda(\omega, A)\, A(\omega)\, x = x, \quad x \in \mathbf{D}. \qquad (2.41)$$

Let $A(\omega) \in \mathbf{L}_s(\Omega, \mathbf{X})$. Then $\bar{R}_\lambda(\omega, A)\, A(\omega)$ is defined as a product of random operators, and it too lies in $\mathbf{L}_s(\Omega, \mathbf{X})$.

Since (2.41) may be extended to the whole of $\mathbf{X}$ by continuity of the terms with respect to x, in the sense of convergence in probability, we have

$$\bar{R}_\lambda(\omega, A) A(\omega) = I - \lambda \bar{R}_\lambda(\omega, A). \tag{2.42}$$

From (2.42) and the inequality $\|\bar{R}_\lambda(\omega, A)\| \leqslant \frac{1}{\lambda}$ we obtain

$$\forall x \in \mathbf{X} \quad \lim_{\lambda\to\infty} \lambda \bar{R}_\lambda(\omega, A) x = x - \lim_{\lambda\to\infty} \bar{R}_\lambda(\omega, A) A(\omega) x = x \tag{2.43}$$

(convergence with probability 1).

Lemma 2. For all $x(\omega) \in \mathbf{X}(\Omega)$ with probability 1,

$$\lim_{\lambda\to\infty} \lambda \bar{R}_\lambda(\omega, A) x(\omega) = x(\omega).$$

Proof. Since $\|\lambda \bar{R}_\lambda(\omega, A)\| \leqslant 1$, it follows that

$$|\lambda \bar{R}_\lambda(\omega, A) x(\omega) - \lambda \bar{R}_\lambda(\omega, A) x_n(\omega)| \leqslant |x(\omega) - x_n(\omega)|.$$

Put $x_m(\omega) = \sum_{k=1}^{m} \xi_k(\omega) e_k$, where $\xi_k(\omega) = (x(\omega), e_k)$.

Then, for any m,

$$\mathbf{P}\{\overline{\lim_{\lambda\to\infty}} |\lambda \bar{R}_\lambda(\omega, A) x(\omega) - x(\omega)| > \varepsilon\} \leqslant$$

$$\leqslant \mathbf{P}\{|x(\omega) - x_m(\omega)| > \varepsilon\} +$$

$$+ \sum_{k=1}^{m} \mathbf{P}\{\overline{\lim_{\lambda\to\infty}} |\lambda \bar{R}_\lambda(\omega, A) e_k - e_k| > 0\}.$$

In view of (2.43) and the convergence of $x_m(\omega)$ to $x(\omega)$ with probability 1, this completes the proof.■

Corollary. For all $x \in \mathbf{X}$ the following holds with probability 1:

$$A(\omega) x = \lim_{\lambda\to\infty} \lambda [x - \lambda \bar{R}_\lambda(\omega, A) x], \tag{2.44}$$

and, in particular, $A(\omega)$ is the strong limit of the sequence of random operators $n(I - n\bar{R}_n(\omega, A))$, which are all in $\mathbf{L}(\Omega, \mathbf{X})$, $(\|n(I - n\bar{R}_n(\omega, A))\|) \leqslant 2n$.

We can utilize (2.44) to define $A(\omega) x(\omega)$ in the case that $x(\omega) \in \mathbf{X}(\Omega)$. Put

$$A(\omega) x(\omega) = \lim_{\lambda\to\infty} \lambda [x(\omega) - \lambda \bar{R}_\lambda(\omega, A) x(\omega)], \tag{2.45}$$

if the limit on the right exists in probability (as $\lambda\bar{R}_\lambda$ is bounded, the expression after the limit sign on the right of (2.45) is defined).

Note that $\bar{R}_\lambda(\omega, A)$ and $\bar{R}_\mu(\omega, A)$ commute, $\lambda > 0$, $\mu > 0$. Hence we can speak of the limit

$$\lim_{\lambda\to\infty} \lambda[\bar{R}_\mu(\omega, A)x - \lambda\bar{R}_\lambda(\omega, A)]\,\bar{R}_\mu(\omega, A) =$$

$$= \lim_{\lambda\to\infty} \lambda[\bar{R}_\mu(\omega, A)x - \bar{R}_\mu(\omega, A)(\lambda\bar{R}_\lambda(\omega, A))\,x] =$$

$$= \bar{R}_\mu(\omega, A)\{\lim_{\lambda\to\infty}[\lambda(x - \lambda\bar{R}_\lambda(\omega, A)\,x)]\}$$

(here we use the fact that $\bar{R}_\mu(\omega, A)$ is bounded, from which it follows that convergence in probability of a sequence $x_n(\omega) \in \mathbf{X}(\Omega)$ to $x(\omega)$ implies convergence in probability of $\bar{R}_\mu(\omega, A)\,x_n(\omega)$ to $\bar{R}_\mu(\omega, A)\,x(\omega)$). Thus, we have

$$A(\omega)\,\bar{R}_\mu(\omega, A)\,x = \bar{R}_\mu(\omega, A)\,A(\omega)\,x \tag{2.46}$$

(the left-hand side is defined through formula (2.45)).

6.4. Equations of Fredholm Type

Let $A(\omega) \in \mathbf{L}_s(\Omega, \mathbf{X})$ be a nonnegative operator and suppose that for some basis $\{e_k\}$ the second inequality of (2.35) is true (the first holds automatically). Consider an equation of the type

$$A(\omega)\,y(\omega) + \lambda y(\omega) = x(\omega). \tag{2.47}$$

The values of $A(\omega)$ on random elements in $\mathbf{X}(\Omega)$ are defined through (2.45), $y(\omega)$ is an unknown r.v. in $\mathbf{X}(\Omega)$ and $x(\omega)$ is a given r.v. We shall show that, under these assumptions, if $\lambda > 0$ then equation (2.47) has a unique solution (up to stochastic equivalence) defined by

$$y(\omega) = \bar{R}_\lambda(\omega, A)\,x(\omega) \tag{2.48}$$

(the operator $\bar{R}_\lambda(\omega, A)$ was defined in the previous subsection). To see that the expression on the right of (2.48) indeed defines a solution of (2.47), we first show that $A(\omega)y(\omega)$ is defined:

$$A(\omega)\bar{R}_\lambda(\omega, A)x(\omega) =$$

$$= \lim_{\mu\to\infty} [\mu(I - \mu\bar{R}_\mu(\omega, A))]\bar{R}_\lambda(\omega, A)x(\omega) =$$

$$= \lim_{\mu\to\infty} [\mu\bar{R}_\lambda(\omega, A) - \mu^2\bar{R}_\mu(\omega, A)\bar{R}_\lambda(\omega, A)]x(\omega).$$

We shall now make use of the relationship that follows from the definition of $\bar{R}_\mu(\omega, A)$ (the resolvent equation):

$$\bar{R}_\lambda(\omega, A) - \bar{R}_\mu(\omega, A) = (\mu - \lambda)\bar{R}_\mu(\omega, A)\bar{R}_\lambda(\omega, A).$$

This implies the equality

$$\bar{R}_\lambda(\omega, A) - \mu\bar{R}_\mu(\omega, A)\bar{R}_\lambda(\omega, A) =$$

$$= \bar{R}_\lambda(\omega, A) - \frac{\mu}{\mu - \lambda}[\bar{R}_\lambda(\omega, A) - \bar{R}_\mu(\omega, A)] =$$

$$= -\frac{\lambda}{\mu - \lambda}\bar{R}_\lambda(\omega, A) + \frac{\mu}{\mu - \lambda}\bar{R}_\mu(\omega, A).$$

Therefore, using Lemma 2, we obtain

$$\lim_{\mu\to\infty} [\mu\bar{R}_\lambda(\omega, A) - \mu^2\bar{R}_\mu(\omega, A)\bar{R}_\lambda(\omega, A)]x(\omega) =$$

$$= -\lambda \lim_{\mu\to\infty} \frac{\mu}{\mu - \lambda}\bar{R}_\lambda(\omega, A)x(\omega) + \lim_{\mu\to\infty} \frac{\mu^2}{\mu - \lambda}\bar{R}_\mu(\omega, A)x(\omega) =$$

$$= -\lambda\bar{R}_\lambda(\omega, A)x(\omega) + x(\omega).$$

We have thus proved that $A(\omega)y(\omega)$ is defined when $y(\omega)$ is defined by formula (2.48), and then

$$A(\omega)y(\omega) = -\lambda\bar{R}_\lambda(\omega, A)x(\omega) + x(\omega) = -\lambda y(\omega) + x(\omega),$$

i.e., equation (2.47) is satisfied.

Now let $y(\omega)$ be an arbitrary solution of (2.47) (in the sense of definition (2.45)). Then, applying the operator $\bar{R}_\lambda(\omega, A)$, to both sides of (2.47), we get

$$\bar{R}_\lambda(\omega, A)A(\omega)y(\omega) + \lambda\bar{R}_\lambda(\omega, A)y(\omega) = \\ = \bar{R}_\lambda(\omega, A)x(\omega). \tag{2.49}$$

Since

$$\bar{R}_\lambda(\omega, A)A(\omega)y(\omega) =$$

$$= \lim_{\mu\to\infty} \{\mu\bar{R}_\lambda(\omega, A)\, y(\omega) - \mu^2 \bar{R}_\lambda(\omega, A)\, \bar{R}_\mu(\omega, A)\, y(\omega)\} =$$

$$= \lim_{\mu\to\infty} \left\{\mu\bar{R}_\lambda(\omega, A) - \frac{\mu^2}{\mu-\lambda}\bar{R}_\lambda(\omega, A) + \frac{\mu^2}{\mu-\lambda}\bar{R}_\mu(\omega, A)\right\} \times$$

$$\times\, y(\omega) = -\lambda\bar{R}_\lambda(\omega, A)\, y(\omega) + y(\omega) \qquad (2.50)$$

(we have again used the resolvent equation and Lemma 2), substitution of this expression in the left-hand side of (2.49) yields (2.48).

6.5. Equations of Evolution Type

Let $A(\omega)$ satisfy the same conditions as in the previous subsection. Consider the equation

$$\frac{d}{dt} x_t(\omega) + A(\omega)\, x_t(\omega) = 0, \quad t \geqslant 0, \qquad (2.51)$$

where $x_t(\omega)$ is an unknown function, $t \geqslant 0$, taking values in $\mathbf{X}(\Omega)$; the initial value $x_0(\omega)$ is assumed given. The operator $A(\omega)$ is again defined for random elements through formula (2.45). By a solution we mean a function $x_t(\omega)$ which has a derivative with respect to t in the sense of convergence in probability, with values in $\mathbf{X}(\Omega)$, and such that application of $A(\omega)$, yields (2.51). We claim that equation (2.51) has a unique solution. Let $x_0 = 0$. Then

$$\frac{d}{dt}(x_t(\omega),\ x_t(\omega)) = -2\,(A(\omega)\, x_t(\omega) : x_t(\omega)) =$$

$$= -2 \lim_{\lambda\to\infty} (\lambda\,(I - \lambda\bar{R}_\lambda(\omega, A))\, x_t(\omega),\ x_t(\omega)) \leqslant 0,$$

since for all $x \in \mathbf{X}$

$$((I - \lambda\bar{R}_\lambda(\omega, A))\, x,\ x) = (x, x) - (\lambda\bar{R}_\lambda(\omega, A)\, x,\ x) \geqslant$$

$$\geqslant (1 - \lambda\|\bar{R}_\lambda(\omega, A)\|)\, |x|^2 \geqslant 0.$$

Therefore, $(x_t(\omega),\ x_t(\omega)) = |x_t(\omega)|^2$ is a decreasing function which vanishes at $t = 0$.

To solve equation (2.51), we apply the operator $\bar{R}_\lambda(\omega, A)$ to both sides. Using (2.50), we see that the equation becomes

$$\frac{d}{dt}\bar{R}_\lambda(\omega, A)\, x_t(\omega) - \lambda \bar{R}_\lambda(\omega, A)\, x_t(\omega) + x_t(\omega) = 0.$$

Multiplying this equality by $e^{-\lambda t}$ and integrating from 0 to ∞ (using the boundedness of $\bar{R}_\lambda$), we obtain

$$\lambda \bar{R}_\lambda(\omega, A) \int_0^\infty e^{-\lambda t} x_t(\omega)\, dt - \bar{R}_\lambda(\omega, A)\, x_0(\omega) -$$

$$- \lambda \bar{R}_\lambda(\omega, A) \int_0^\infty e^{-\lambda t} x_t(\omega)\, dt + \int_0^\infty e^{-\lambda t} x_t(\omega)\, dt = 0.$$

Thus, the Laplace transform of the solution is defined by

$$\int_0^\infty e^{-\lambda t} x_t(\omega)\, dt = \bar{R}_\lambda(\omega, A)\, x_0(\omega). \tag{2.52}$$

The existence of the Laplace transform follows from the proven fact that $| x_t(\omega) |^2$, is bounded. The verification that $\bar{R}_\lambda(\omega, A)\, x_0(\omega)$ is indeed the Laplace transform of a differentiable function must be done separately in each individual case. If there exists a differentiable function $z_t(\omega)$, such that

$$\int_0^\infty e^{-\lambda t} z_t(\omega)\, dt = \bar{R}_\lambda(\omega, A)\, x_0(\omega),$$

then it is readily seen that for all $\lambda > 0$

$$\frac{d}{dt}\bar{R}_\lambda(\omega, A)\, z_t(\omega) - \lambda \bar{R}_\lambda(\omega, A)\, z_t(\omega) + z_t(\omega) = 0.$$

Hence

$$\lambda \bar{R}_\lambda(\omega, A) \frac{d}{dt} z_t(\omega) + \lambda [z_t(\omega) - \lambda \bar{R}_\lambda(\omega, A)\, z_t(\omega)] = 0.$$

Letting $\lambda \to \infty$ and using Lemma 2 and formula (2.45), we see that (2.51) holds.

One can also point out functions $x_0(\omega)$ for which the equation is solvable. Let $x_0(\omega) = (R_1(\omega, A))^3 x$ (the cube of the operator). Then

$$\overline{R}_\lambda(\omega, A)\, x_0(\omega) = \frac{1}{\lambda}\, x_0(\omega) - \frac{1}{\lambda^2}\, A(\omega)\, x_0(\omega) +$$

$$+ \frac{1}{\lambda^3}\, A^2(\omega)\, x_0(\omega) - \frac{1}{\lambda^3}\, A^3(\omega)\, \overline{R}_\lambda(\omega, A)\, x_0(\omega).$$

The expressions on the right are defined through the equalities

$$\begin{aligned}
&A(\omega)\, x_0(\omega) = A(\omega)\, \overline{R}_1(\omega, A)\, (\overline{R}_1(\omega, A))^2\, x = \\
&= (I - \overline{R}_1(\omega, A))\, (\overline{R}_1(\omega, A))^2\, x, \\
&A^2(\omega)\, x_0(\omega) = A(\omega)\, (I - \overline{R}_1(\omega, A))\, (\overline{R}_1(\omega, A))^2\, x = \\
&= (I - \overline{R}_1(\omega, A))^2\, \overline{R}_1(\omega, A)\, x, \\
&A^3(\omega)\, \overline{R}_\lambda(\omega, A)\, x_0(\omega) = \overline{R}_\lambda(\omega, A)\, (I - \overline{R}_1(\omega, A))^3\, x,
\end{aligned}$$

which are derived with the help of (2.42), (2.46). Thus,

$$\overline{R}_\lambda(\omega, A)\, x_0(\omega) = \int_0^\infty e^{-\lambda t} x_0(\omega)\, dt - \int_0^\infty t e^{-\lambda t} A(\omega)\, x_0(\omega)\, dt +$$

$$+ \frac{1}{2} \int_0^\infty e^{-\lambda t} t^2 A^2(\omega)\, x_0(\omega)\, dt - \frac{1}{\lambda^3}\, \overline{R}_\lambda(\omega, A)\, A^3(\omega)\, x_0(\omega).$$

The first three terms constitute the Laplace transform of the differentiable function

$$x_0(\omega) - tA(\omega)\, x_0(\omega) + \frac{t^2}{2}\, A^2(\omega)\, x_0(\omega).$$

Using the inverse Laplace transform, and the fact that $\overline{R}_\lambda(\omega) A^3(\omega) x_0(\omega)$ is bounded when $\operatorname{Re}\lambda > \delta > 0$, we find that the last term is the Laplace transform of the function

$$\frac{1}{2\pi i} \int_{\operatorname{Re}\lambda = \delta} e^{\lambda t}\, \frac{1}{\lambda^3}\, \overline{R}_\lambda(\omega, A)\, A^3(\omega)\, x_0(x)\, d\lambda,$$

which is also differentiable.

CHAPTER 3

OPERATOR-VALUED MARTINGALES

7. Operator-Valued Martingale Sequences

7.1. Weak Operator-valued Martingale

Let $A_t(\omega)$, $t \in T \subset R$ be operators in $\mathbf{L}_w(\Omega, \mathbf{X})$ and suppose that for all $t \in T$ we have a σ-algebra $\mathfrak{f}_t$, $\mathfrak{f}_t \subset \mathfrak{S}$ (where $\{\Omega, \mathfrak{S}, \mathbf{P}\}$ is the original probability space) such that the following conditions hold:

(1) $\mathfrak{f}_{t_1} \subset \mathfrak{f}_{t_2}$ whenever $t_1 < t_2$;

(2) $(A_t(\omega)\, x, y)$ is $\mathfrak{f}_t$ -measurable for all $x, y \in \mathbf{X}$.

The family $A_t(\omega)$ is called a weak operator-valued martingale relative to $\mathfrak{f}_t$ if, in addition to (1) and (2), the following conditions hold:

(3) $\mathbf{M}\,|(A_t(\omega)\, x, y)| < \infty$ for $t \in T$; $x, y \in \mathbf{X}$;

(4) for any $t < s$ $x, y \in \mathbf{X}$

$$\mathbf{M}\,\{(A_s(\omega)\, x, y)/\mathfrak{f}_t\} = (A_t(\omega)\, x, y). \tag{3.1}$$

If $A_t(\omega)$ is a weak martingale (the qualification "operator--valued" will be omitted whenever we are dealing with random operators) relative to a filtration of σ-algebras $\mathfrak{f}_t$ (i.e., a family of σ-algebras satisfying condition (1) above), then it is a weak martingale relative to the filtration $\mathfrak{f}_t^*$, where $\mathfrak{f}_t^*$ is the smallest σ-algebra relative to which $(A_s(\omega)\, x, y)$, $s \in T$, $s \leqslant t$, $x, y \in \mathbf{X}$, are measurable. Since the filtration $\mathfrak{f}_t^*$ is defined by the original family $\{A_t(\omega), t \in T\}$, a weak martingale relative to $\mathfrak{f}_t^*$ will be called simply a weak martingale.

In this section we shall deal with the case in which $T = \{1, 2, \ldots\}$ is the set of natural numbers; the term "martingale sequences" is sometimes used in that case. Obviously, if $\{A_n(\omega)\}$ is a weak martingale, then $(A_n(\omega)\, x, y)$ is a (real--valued) martingale relative to the same filtration $\mathfrak{f}_t^*$, underlying the martingale $\{A_n(\omega)\}$. This simple fact makes it possible to apply known results concerning real-valued martingales to operator-valued martingales. As an example, we consider the following theorem on the existence of limits.

Theorem 1. Let $\{A_n(\omega),\ n = 1, 2, \ldots\}$ be a weak operator--valued martingale for which $\sup_n \mathbf{M}\,|(A_n(\omega)x, y)| < \infty$ for all $x, y \in \mathbf{X}$. Suppose that, in addition, there is a dense set $\mathbf{D} \subset \mathbf{X}$ such that the r.v.'s $(A_n(\omega)x, y)$ are uniformly (in n) integrable for all $x, y \in \mathbf{D}$. Then

$$\mathbf{M}(A(\omega)x, y) = \lim_{n\to\infty} \mathbf{M}(A_n(\omega)x, y); \tag{3.2}$$

$$(A_n(\omega)x, y) = \mathbf{M}\{(A(\omega)x, y)/\mathfrak{f}_n\}. \tag{3.3}$$

Proof. The first assertion of the theorem follows from the theorem on limits of real-valued martingales [8, p. 82] and from Theorem 1, § 3. The theorem on the existence of closures of real-valued martingales [8, p. 84] implies that (3.2) and (3.3) are equivalent and that they hold for $x, y \in \mathbf{D}$. Put

$$(\hat{A}_n(\omega)x, y) = \mathbf{M}\{(A(\omega)x, y)/\mathfrak{f}_n\}, \tag{3.4}$$

where $A(\omega)$ is the weak limit of $A_n(\omega)$. It is clear that this conditional expectation exists and satisfies condition W1) (bilinearity in x and y). To see that $\hat{A}_n(\omega) \in \mathbf{L}_w(\Omega, \mathbf{X})$, we must show that the right-hand side of (3.4) is bounded in probability for $|x| < 1$, $|y| < 1$; this follows from the inequality

$$\mathbf{M}\,|(\hat{A}_n(\omega)x, y)| \leqslant \mathbf{M}\mathbf{M}\{|(A(\omega)x, y)|/\mathfrak{f}_n\} = \mathbf{M}\,|(A(\omega)x, y)|$$

and from Lemma, § 2. We now note that for $x, y \in \mathbf{D}$

$$(\hat{A}_n(\omega)x, y) = (A_n(\omega)x, y),$$

with probability 1, so that $\hat{A}_n(\omega) = A_n(\omega)$.

7.2. Strong Operator-valued Martingales

Let $A_t(\omega)$, $s \in T \subset \mathbf{R}$ belong to $\mathbf{L}_s(\Omega, \mathbf{X})$ and suppose that it is a weak martingale relative to a filtration $\mathfrak{f}_t$. If $\mathbf{M}\,|A_t(\omega)$ $(\omega)x| < \infty$, for all $x \in \mathbf{X}$, $t \in T$, then $\{A_t(\omega), t \in T\}$ is called a strong operator-valued martingale relative to $\mathfrak{f}_t$. We now consider existence conditions for limits of strong operator--valued martingales.

Theorem 2. Let $\{A_n(\omega),\ n = 1, 2, \ldots\}$ be a strong martingale such that

$$\sup_n \mathbf{M} \,|\, A_n(\omega)\, x \,| < \infty, \quad x \in \mathbf{X}. \tag{3.5}$$

Then the strong limit

$$A(\omega)\, x = \lim_{n\to\infty} A_n(\omega)\, x, \quad x \in \mathbf{X} \tag{3.6}$$

exists with probability 1 if and only if there exists an absolutely continuous symmetric positive operator S such that

$$\overline{\lim_{\alpha\to\infty}} \sup_n \mathbf{P}\{|S^{-1}A_n(\omega)\, x| > \alpha\} = 0 \quad \forall x \in \mathbf{X}. \tag{3.7}$$

Proof. Necessity follows from Theorem 2 in [8, p. 379]; indeed, the existence of the limit on the right of (3.7) implies that the measures $\mu_n(\cdot)$, defined as the distributions of the r.v.'s $A_n(\omega)x$, are weakly convergent, and therefore, by the above-mentioned theorem, there exists an operator S such that

$$\lim_{\alpha\to\infty} \sup_n \mu_n(\{x\colon |S^{-1}x| > \alpha\}) = 0,$$

and this is equivalent to (3.7).

Sufficiency. Theorem 1 implies the existence, for all $x, y \in \mathbf{X}$ of the limit

$$\lim_{n\to\infty} (A_n(\omega)\, x,\, y) = (A(\omega)\, x,\, y),$$

where $A(\omega) \in \mathbf{L}_w(\Omega, \mathbf{X})$, since $\mathbf{M}\,|\,(A_n(\omega)x, \ y)\,| \leqslant |\,y\,|\,\mathbf{M}\,|\,A_n(\omega)x\,|$ and inequality (3.5) is valid. We claim that $A(\omega) \in \mathbf{L}_s(\Omega, \mathbf{X})$. To prove this it will suffice to show that $A_n(\omega)\, x$ is stongly convergent in probability. It follows from (3.7) that for any $\varepsilon > 0$ there exists α such that

$$\mathbf{P}\{|S^{-1}A_n(\omega)\, x| > \alpha\} < \varepsilon \quad \text{for all } n.$$

Since $\{y : |\, S^{-1}y \,| \leqslant \alpha\}$ is a closed compact set, there exists a finite-dimensional subspace $\mathbf{N} \subset \mathbf{X}$, constituting a δ-net in it. Let Q be the projection operator onto $\mathbf{N}$. Then

$$\mathbf{P}\{|\, A_n(\omega)\, x - A_m(\omega)\, x \,| > 3\delta\} \leqslant$$

$$\leqslant \mathbf{P}\{|S^{-1}A_n(\omega)\, x| > \alpha\} + \mathbf{P}\{|S^{-1}A_m(\omega)\, x| > \alpha\} +$$

$$+ \mathbf{P}\{|\, QA_n(\omega)\, x - QA_m(\omega)\, x \,| > \delta\},$$

since for $z \in \{y : |S^{-1}y| \leqslant \alpha\}$, $|z - Qz| < \delta$. Let $f_1, \dots f_l$ be a basis in $\mathbf{N}$. Then

$$\lim_{n,m\to\infty} \mathbf{P}\{|QA_n(\omega)x - QA_m(\omega)x| > \delta\} \leqslant$$

$$\leqslant \overline{\lim_{n,m\to\infty}} \sum_{i=1}^{l} \mathbf{P}\left\{|(A_n(\omega)x, f_i) - (A_m(\omega)x, f_i)| > \frac{\delta}{l}\right\} = 0.$$

Thus

$$\overline{\lim_{n,m\to\infty}} \mathbf{P}\{|A_n(\omega)x - A_m(\omega)x| > 3\delta\} \leqslant 2\varepsilon.$$

It follows from this that $A_n(\omega)x$ is strongly convergent in probability to some limit. Since $(A_n(\omega)x, y)$ converges to $(A(\omega)x, y)$ for all y, it follows that

$$\Sigma(A_n(\omega)x, e_k)e_k \to \Sigma(A(\omega)x, e_k)e_k$$

in probability, and moreover the series on the right is convergent and so $A(\omega) \in \mathbf{L}_s(\Omega, \mathbf{X})$ and $A_n(\omega)x \to A(\omega)x$ in probability for all x. Hence $|A_n(\omega)x| \to |A(\omega)\mathbf{X}|$ in probability. Note, moreover, that $|A_n(\omega)x|$ is a (real-valued) submartingale:

$$\mathbf{M}(|A_n(\omega)x|/\mathfrak{f}_{n-1}) = \mathbf{M}(\sup_k (A_n(\omega)x, y_k)/\mathfrak{f}_{n-1}) \geqslant$$

$$\geqslant \sup_k (A_{n-1}(\omega)x, y_k) = |A_{n-1}(\omega)x|,$$

where $\{y_k\}$ is a sequence of vectors in $\mathbf{X}$, which is dense on the unit sphere $\{x : |x| = 1\}$. Condition (3.5), Theorem 1 and the corollary in [8, pp. 81-82] now imply that the limit $\lim_{n\to\infty} |A_n(\omega)x|$, exists with probability 1; this is clearly the same as the limit in probability.

Therefore $\lim_{n\to\infty} |A_n(\omega)x| = |A(\omega)x|$. Thus, there exists $\Lambda \subset \mathfrak{S}$, such that $\mathbf{P}(\Lambda) = 1$ and a countable subset $\mathbf{D}$ of $\mathbf{X}$, such that for all $y \in \mathbf{D}$ and $\omega \in \Lambda$

$$\lim_{n\to\infty} (A_n(\omega)x, y) = (A(\omega)x, y), \quad \lim_{n\to\infty} |A_n(\omega)x| = |A(\omega)x|.$$

It follows from these relationships that for $\omega \in \Lambda$

$$\lim A_n(\omega)x = A(\omega)x.$$

Remark. Under the assumptions of the theorem, if there exists a dense subset $\mathbf{D}$ of $\mathbf{X}$, such that the sequence $(A_n(\omega)x, y)$ is integrable uniformly in n, then

$$\mathbf{M}A(\omega)x = \lim_{n\to\infty} \mathbf{M}A_n(\omega)x; \tag{3.8}$$

$$A_n(\omega)x = \mathbf{M}[A(\omega)x/\mathfrak{f}_n]. \tag{3.9}$$

This assertion follows directly from Theorem 1.

More conveniently checked conditions for the existence of the limit can be obtained if one assumes the existence of second moments.

Theorem 3. Let $\{A_n(\omega), n = 1, 2, \ldots\}$ be a strong martingale such that

$$\forall x \in \mathbf{X} \quad \sup_n \mathbf{M}|A_n(\omega)x|^2 < \infty. \tag{3.10}$$

Then the strong limit $\lim\limits_{n\to\infty} A_n(\omega)x = A(\omega)x$, $x \in \mathbf{X}$ exists with probability 1, and inequalities (3.8) and (3.9) are valid.

Proof. The existence of the limits $\lim\limits_{n\to\infty}(A_n(\omega)x, y) = (A(\omega)x, y)$ follows from Theorem 1. We now show that $A_n(\omega)x$ is convergent in mean square to some limit. For $m > n$,

$$\mathbf{M}(A_n(\omega)x, A_m(\omega)x) = \mathbf{M}\,\mathbf{M}\{(A_n(\omega)x, A_m(\omega)x)/\mathfrak{f}_n\} =$$

$$= \mathbf{M}(A_n(\omega)x, \mathbf{M}(A_m(\omega)x/\mathfrak{f}_n)) = \mathbf{M}(A_n(\omega)x, A_n(\omega)x).$$

Therefore,

$$\lim_{n,m\to\infty} \mathbf{M}(A_n(\omega)x - A_m(\omega)x, A_n(\omega)x - A_m(\omega)x) =$$

$$= \lim_{n,m\to\infty} \mathbf{M}[(A_n(\omega)x - A_n(\omega)x) - 2(A_n(\omega)x, A_m(\omega)x) +$$

$$+ (A_m(\omega)x, A_m(\omega)x)] = \lim_{n,m\to\infty}[\mathbf{M}(A_n(\omega)x, A_n(\omega)x) -$$

$$- \mathbf{M}(A_m(\omega)x, A_m(\omega)x)] = 0.$$

Thus $A_n(\omega)x$ is strongly convergent in probability to $A(\omega)x$. The rest of the proof proceeds in a manner similar to that of Theorem 2.

Remark. Let $x_n(\omega)$ be a sequence of r.v.'s in $\mathbf{X}$. It is said to be a martingale if $\mathbf{M}|x_n(\omega)| < \infty$ and

$\mathbf{M}(x_m(\omega)/\mathfrak{f}_n) = x_n(\omega)$ for $m > n$, where $\mathfrak{f}_n$ is the σ-algebra generated by the r.v.'s $x_1(\omega), \ldots, x_n(\omega)$. It follows from Theorem 3 that a martingale $x_n(\omega)$ has a limit if $\sup_n \mathbf{M}|x_n(\omega)|^2 < \infty$. To verify this it suffices to apply the theorem to the sequence of operators $\langle a \circ x_n(\omega)\rangle$, where $a \in \mathbf{X}$.

7.3. Operator-valued Martingale

A sequence of random operators $A_n(\omega)$ in $\mathbf{L}(\Omega, \mathbf{X})$ is called an operator-valued martingale relative to a filtration of σ-algebras $\mathfrak{f}_n$, if:

1. $\forall x, y \in \mathbf{X}$, $(A_n(\omega)x, y)$ is a real-valued martingale relative to $\mathfrak{f}_n$;
2. $\mathbf{M}\|A_n(\omega)\| < \infty$.

We are interested in conditions under which an operator-valued martingale has a limit in operator norm.

Theorem 4. Let $A_n(\omega)$ be an operator-valued martingale satisfying the following conditions:

(1) the sequence $\|A_n(\omega)\|$ is uniformly integrable;

(2) for any ε > 0 there exists a finite-dimensional subspace $\mathbf{N}$ such that for all n

$$\mathbf{P}\{\|P_{\mathbf{N}}A_n(\omega)P_{\mathbf{N}} - A_n(\omega)\| > \varepsilon\} < \varepsilon, \tag{3.11}$$

where $P_{\mathbf{N}}$ is the projection operator onto $\mathbf{N}$. Then there exists an operator $A(\omega) \in \mathbf{L}(\Omega, \mathbf{X})$ such that

$$\mathbf{P}\{\lim_{n\to\infty}\|A_n(\omega) - A(\omega)\| = 0\} = 1.$$

Proof. The operator $A(\omega)$, should naturally be the same as the weak limit of the sequence $A_n(\omega)$, whose existence follows from Theorem 1 since

$$\sup_n \mathbf{M}|(A_n(\omega)x, y)| \leqslant |x||y| \sup_n \mathbf{M}\|A_n(\omega)\|.$$

We shall show that $A_n(\omega)$ converges in operator norm in probability to some limit. Note that $P_{\mathbf{N}}A_n(\omega)P_{\mathbf{N}}$ is also a martingale relative to $\mathfrak{f}_n$, since

$$\mathbf{M}[(P_{\mathbf{N}}A_n(\omega)P_{\mathbf{N}}x, y)/\mathfrak{f}_{n-1}] = \mathbf{M}[(A_n(\omega)P_{\mathbf{N}}x, P_{\mathbf{N}}y)/\mathfrak{f}_{n-1}] =$$

$$= (A_{n-1}(\omega)P_{\mathbf{N}}x, P_{\mathbf{N}}y) = (P_{\mathbf{N}}A_{n-1}(\omega)P_{\mathbf{N}}x, y).$$

Moreover, there exists a weak limit

$$\lim (P_N A_n(\omega) P_N x, y) = (P_N A(\omega) P_N x, y).$$

If $\{e_1, \ldots, e_l\}$ is a basis in $\mathbf{N}$, then

$$\| P_N A_n(\omega) P_N - P_N A(\omega) P_N \| \leqslant \sqrt{\sum_{k,j=1}^{l} ((A_n(\omega) - A(\omega)) e_k, e_j)^2};$$
$$\lim_{n\to\infty} \| P_N A_n(\omega) P_N - P_N A(\omega) P_N \| = 0, \qquad (3.12)$$

for any $\mathbf{N}$. Therefore, for $\delta < \varepsilon$,

$$\overline{\lim_{\substack{n\to\infty\\ m\to\infty}}} \mathbf{P}\{\| A_n(\omega) - A_m(\omega)\| > \varepsilon\} \leqslant$$

$$\leqslant \overline{\lim_{\substack{n\to\infty\\ m\to\infty}}} \Big[\mathbf{P}\{\| P_N A_n(\omega) P_N - P_N A(\omega) P_N \| >$$

$$> \varepsilon - \delta\} + \mathbf{P}\left\{\| P_N A_n(\omega) P_N - A_n(\omega)\| > \frac{\delta}{2}\right\} +$$

$$+ \mathbf{P}\left\{\| P_N A_n(\omega) P_N - A_m(\omega)\| > \frac{\delta}{2}\right\}\Big] \leqslant \delta$$

(the first term on the right vanishes thanks to (3.12)). It follows from the last inequality that $A_n(\omega)$ is convergent in norm in probability. The limit is precisely $A(\omega)$. It is obvious that $P_N A_n(\omega) P_N - A_n(\omega)$ is convergent in norm in probability to $P_N A(\omega) P_N - A(\omega)$. Hence, by (3.11),

$$\mathbf{P}\{\| P_N A(\omega) P_N - A(\omega)\| > \varepsilon\} < \varepsilon. \qquad (3.13)$$

Choose a sequence $\varepsilon_k \downarrow 0$ so that $\Sigma \varepsilon_k < \infty$. Let $\mathbf{N}_k$ be finite--dimensional subspaces satisfying condition (2) of the theorem for ε_k. It follows from (3.13) that

$$\mathbf{P}\{\lim_{k\to\infty} \| P_{N_k} A(\omega) P_{N_k} - A(\omega)\| = 0\} = 1. \qquad (3.14)$$

Note that $\| A_n(\omega)\|$ converges in probability to $\| A(\omega)\|$, and therefore, by condition 1 of the theorem, $\mathbf{M}\|A(\omega)\| < \infty$. Since $\| P_{N_k} A(\omega) P_{N_k} - A(\omega)\| \leqslant 2\| A(\omega)\|$, it follows from (3.14) by the Lebesgue theorem that

$$\lim_{k\to\infty} \mathbf{M}\| P_{N_k} A(\omega) P_{N_k} - A(\omega)\| = 0. \qquad (3.15)$$

The sequence $\|P_{\mathbf{N}}A_n(\omega)P_{\mathbf{N}} - A_n(\omega)\|$ is a submartingale, whatever $\mathbf{N}$ may be, since by taking a sequence $\{x_k\}$ dense on the unit sphere we obtain

$$\|P_{\mathbf{N}}A_n(\omega)P_{\mathbf{N}} - A_n(\omega)\| = \sup_{k,j}((P_{\mathbf{N}}A_n(\omega)P_{\mathbf{N}} - A_n(\omega))x_k, x_j)$$

(the term on the left is the supremum of a countable set of martingales). It follows from the inequality

$$\|P_{\mathbf{N}}A_n(\omega)P_{\mathbf{N}} - A_n(\omega)\| \leqslant 2\|A_n(\omega)\|$$

that this submartingale is uniformly integrable and so

$$\mathbf{M}\|P_{\mathbf{N}}A_n(\omega)P_{\mathbf{N}} - A_n(\omega)\| \uparrow \mathbf{M}\|P_{\mathbf{N}}A(\omega)P_{\mathbf{N}} - A(\omega)\|$$

(by what we have proved, $\|P_{\mathbf{N}}A_n(\omega)P_{\mathbf{N}} - A_n(\omega)\|$ is convergent in probability to $\|P_{\mathbf{N}}A(\omega)P_{\mathbf{N}} - A(\omega)\|$; hence it is convergent to the same limit with probability 1). Using inequality (25) of [8, p. 81], we can assert that

$$\mathbf{P}\{\sup_n\|P_{\mathbf{N}}A_n(\omega)P_{\mathbf{N}} - A_n(\omega)\| \geqslant \alpha\} \leqslant$$
$$\leqslant \frac{1}{\alpha}\mathbf{M}\|P_{\mathbf{N}}A(\omega)P_{\mathbf{N}} - A(\omega)\|. \tag{3.16}$$

Further,

$$\mathbf{P}\{\overline{\lim_{n\to\infty}}\|A_n(\omega) - A(\omega)\| > \delta\} \leqslant$$
$$\leqslant \mathbf{P}\left\{\sup_n\|P_{\mathbf{N}_k}A_n(\omega)P_{\mathbf{N}_k} - A_n(\omega)\| > \frac{\delta}{3}\right\} +$$
$$+ \mathbf{P}\left\{\|P_{\mathbf{N}_k}A(\omega)P_{\mathbf{N}_k} - A(\omega)\| > \frac{\delta}{3}\right\} +$$
$$+ \mathbf{P}\left\{\overline{\lim_{n\to\infty}}\|P_{\mathbf{N}_k}A_n(\omega)P_{\mathbf{N}_k} - P_{\mathbf{N}_k}A(\omega)P_{\mathbf{N}_k}\| > \frac{\delta}{3}\right\}.$$

The last probability on the right is zero because of (3.12); the other two may be made arbitrarily small by suitable choice of k - this follows from (3.16) and (3.15).■

Remark 1. By the assumptions of the theorem, $(A_n(\omega)x, y)$ is uniformly integrable; hence, by Theorem 1,

$$\mathbf{M}(A(\omega)/\mathfrak{f}_n) = A_n(\omega).$$

Remark 2. Let $\mathbf{L}^c(\mathbf{X})$ denote the set of all completely continuous linear operators. The set $\mathbf{L}^c(\mathbf{X})$ with the operator norm and the natural operations of addition and multiplication by real numbers is a separable Banach space. The set of operators $P_{\mathbf{N}} A P_{\mathbf{N}}$, where $\mathbf{N}$ is a finite-dimensional subspace of $\mathbf{X}$, is everywhere dense in $\mathbf{L}^o(\mathbf{X})$ It is readily established that for any compact subset $R \subset \mathbf{L}^c(\mathbf{X})$ and any $\varepsilon > 0$ there exists a finite-dimensional subspace $\mathbf{N}$, such that

$$\sup_{A \in R} \|P_{\mathbf{N}} A P_{\mathbf{N}} - A\| < \varepsilon. \tag{3.17}$$

Conversely, if R is a bounded closed set such that for any $\varepsilon > 0$ there exists a finite-dimensional $\mathbf{N}$, satisfying (3.17), then R is compact. Hence it follows that the conditions (a) $A(\omega) \in \mathbf{L}(\Omega, \mathbf{X})$, (b) for any $\varepsilon > 0$ there exists a finite--dimensional subspace $\mathbf{N}$, such that

$$\mathbf{P}\{\|P_{\mathbf{N}} A(\omega) P_{\mathbf{N}} - A(\omega)\| > \varepsilon\} < \varepsilon$$

are necessary and sufficient for the truth of $\mathbf{P}\{A(\omega) \in \mathbf{L}^c(\mathbf{X})\} = 1$. Thus, the martingales considered in Theorem 4 assume values in $\mathbf{L}^c(\mathbf{X})$. Then condition (2) of the theorem is also necessary for convergence. Indeed, by the weak convergence theorem for measures in a complete separable metric space [8, p. 379], for any $\varepsilon > 0$ there exists a compact set $R \subset \mathbf{L}^c(\mathbf{X})$, such that $\mathbf{P}\{A_n(\omega) \in R\} \geqslant 1 - \varepsilon$ for all n. But then, by (3.17),

$$\mathbf{P}\{\|P_{\mathbf{N}} A_n(\omega) P_{\mathbf{N}} - A_n(\omega)\| < \varepsilon\} \geqslant 1 - \varepsilon.$$

Following is an example of an operator-valued martingale for which condition (1) of the theorem is satisfied, but $\lim_{n\to\infty} A_n(\omega)$ does not exists in the operator norm.

Example. Let $\{e_k\}$ be an orthonormal basis in $\mathbf{X}$, Q_k the projection operator onto e_k, ξ_k a sequence of independent r.v.'s such that $\mathbf{M}\xi_k = 0$. Then the sequence

$$A_n(\omega) = \sum_{k=1}^{n} \xi_k Q_k$$

is an operator-valued martingale, $\|A_n(\omega)\| = \sup_{k \leqslant n} |\xi_k|$. Thus condition (1) of the theorem will be satisfied if

$$\mathbf{M} \sup_k |\xi_k| < \infty.$$

In that case $A(\omega) = \sum_{k=1}^{\infty} \xi_k Q_k$ is a random operator, but

$$\| A(\omega) - A_n(\omega) \| = \sup_{k>n} |\xi_k|,$$

and it is easy to choose the sequence ξ_k, in such a way that $\sup_{k>n} |\xi_k|$ does not tend to zero (e.g., take the ξ_k bounded and identically distributed).

8. Convergence of Infinite Products of Independent Random Operators

8.1. Infinite Products as Martingales

Let $B_n(\omega)$ be a sequence of independent operators in $\mathbf{L}_s(\Omega, \mathbf{X})$. We are interested in conditions for the convergence of the infinite product

$$\prod_{k=1}^{\infty} (I + B_k(\omega)), \tag{3.18}$$

i.e., for the existence of a limit of the products

$$\prod_{k=1}^{n} (I + B_k(\omega)).$$

The first thing to mention is that this product is defined and an element of $\mathbf{L}_s(\Omega, \mathbf{X})$. This follows from the following proposition.

Lemma. If $A(\omega)$ and $B(\omega)$ are in $\mathbf{L}_s(\Omega, \mathbf{X})$ and independent, then $A(\omega)B(\omega)$ is defined and belongs to $\mathbf{L}_s(\Omega, \mathbf{X})$.

Proof. Let $\mathfrak{f}_B$ denote the σ-algebra generated by the operator $B(\omega)$. For any $z \in \mathbf{X}$ the series

$$\sum_{k=1}^{\infty} (A(\omega) z, e_k)^2 = |A(\omega) z|^2$$

is convergent in probability. Thus, for any z and any $\varepsilon > 0$,

$$\lim_{n\to\infty} \sup_m \mathbf{P}\{|\sum_{k=n}^{n+m} (A(\omega) z, e_k)^2| > \varepsilon\} = 0. \qquad (3.19)$$

Since $B(\omega) x \in \mathbf{X}$ with probability 1,

$$\mathbf{P}\{\sum_{k=n}^{n+m} (A(\omega) B(\omega) x, e_k)^2 > \varepsilon/\mathfrak{f}_B\} =$$

$$= \mathbf{P}\{\sum_{k=n}^{n+m} (A(\omega) z, e_k)^2 > \varepsilon\}|_{z=B(\omega)x}$$

and since $A(\omega)$ and $B(\omega)$, are independent,

$$\mathbf{P}\{\sum_{k=n}^{n+m} (A(\omega) B(\omega) x, e_k)^2 > \varepsilon\} =$$

$$= \int \mathbf{P}\{\sum_{k=n}^{n+m} (A(\omega) z, e_k)^2 > \varepsilon\} \mathbf{P}\{B(\omega) x \in dz\}. \qquad (3.20)$$

The proof of the lemma now follows from (3.19) and (3.20) via the Lebesgue convergence theorem. ∎

Consider the sequence

$$A_n(\omega) = \prod_{k=1}^{n} (I + B_k(\omega)).$$

By the lemma, $A_n(\omega) \in \mathbf{L}_s(\Omega, \mathbf{X})$. We claim that if $\mathbf{M}|B_k(\omega) x| < \infty$, $\mathbf{M}B_k(\omega) x = 0$ for all x, then $A_n(\omega)$ is a strong operator-valued martingale. Indeed, from the lemma of § 2 it follows that there are constants β_k, such that

$$\mathbf{M}|B_k(\omega) x| \leqslant \beta_k |x|.$$

Next we note that if $A(\omega)$ and $B(\omega)$, are independent operators for which $\mathbf{M}|A(\omega) x| \leqslant \alpha |x|$, $\mathbf{M}|B(\omega) x| \leqslant \beta x$ then

$$\mathbf{M}(|A(\omega) B(\omega) x|) = \mathbf{M}[\mathbf{M}|A(\omega) z||_{z=B(\omega)x}] \leqslant$$

$$\leqslant \alpha \mathbf{M}|B(\omega) x| \leqslant \alpha\beta |x|.$$

Since

$$\mathbf{M}|x + B_k(\omega) x| \leqslant |x| + \mathbf{M}|B_k(\omega)| x \leqslant (1 + \beta_k)|x|,$$

we have

$$\mathbf{M}\,|A_n(\omega)x| \leqslant \prod_{k=1}^{n}(1+\beta_k)\,|x|. \tag{3.21}$$

If $\mathfrak{f}_n$ denotes the σ-algebra generated by $B_k(\omega)$, $k = 1, \ldots, n$, then $B_{n+1}(\omega)$ is independent of $\mathfrak{f}_n$, and $A_n(\omega)$ is measurable relative to $\mathfrak{f}_n$; hence

$$\mathbf{M}\,[(A_{n+1}(\omega)x, y)/\mathfrak{f}_n] = (A_n(\omega)x, y) +$$

$$+\, \mathbf{M}\,[(A_n(\omega)B_{n+1}(\omega)x, y)/\mathfrak{f}_n].$$

We shall show that the second term on the right vanishes. Let $\xi(\omega)$ be an arbitrary $\mathfrak{f}_n$ -measurable bounded r.v. Then

$$\mathbf{M}\,(\xi(\omega)A_n(\omega)B_{n+1}(\omega)x, y) =$$

$$= \mathbf{MM}\,((\xi(\omega)A_n(\omega)B_{n+1}(\omega)x, y)/\mathfrak{f}_{B_{n+1}}) =$$

$$= \mathbf{M}\,\{\mathbf{M}\,(\xi(\omega)A_n(\omega)z, y)_{z=B_{n+1}(\omega)x}\},$$

where $\mathfrak{f}_{B_{n+1}}$ is the σ-algebra generated by $B_{n+1}(\omega)$. Since $\mathbf{M}\,|(\xi(\omega)A_n(\omega)z, y)| < \infty$ for any $z, y \in \mathbf{X}$, it follows by the lemma of § 2 that there exists a bounded operator S for which $\mathbf{M}\,(\xi(\omega)A_n(\omega)z, y) = (Sz, y)$. Therefore,

$$\mathbf{M}\,\{\mathbf{M}\,(\xi(\omega)A_n(\omega)z, y)_{z=B_{n+1}(\omega)x}\} =$$

$$= \mathbf{M}\,(SB_{n+1}(\omega)x, y) = \mathbf{M}\,(B_{n+1}(\omega)x, S^*y = 0.$$

We have thereby proved that

$$\mathbf{M}\,[(A_{n+1}(\omega)x, y)/\mathfrak{f}_n] = (A_n(\omega)x, y).$$

It follows from (3.21) that $A_n(\omega)$ is a strong operator-valued martingale.

We can now use martingale convergence theorems to prove the convergence of an infinite product. We present one such general theorem.

Theorem 1. Let $\mathbf{M}B_k(\omega)x = 0$ and assume that for all $x \in \mathbf{X}$, $\mathbf{M}\,|B_k(\omega)x|^2 < \infty$. Let V_k denote nonnegative bounded symmetric operators such that $\mathbf{M}\,|B_k(\omega)x|^2 = (V_k x, x)$ if the series $\sum_{k=1}^{\infty}(V_k x, x)$ is convergent for all $x \in \mathbf{X}$ and there exists l

such that

$$\left\| \sum_{k=l}^{\infty} V_k \right\| < 1,$$

then $\prod_{k=1}^{n} (I + B_k(\omega))\, x$ has a limit with probability 1, for all $x \in \mathbf{X}$.

<u>Proof</u>. Since $A_n(\omega) = \prod_{k=1}^{n} (I + B_k(\omega))$ is a strong martingale, it follows in view of Theorem 3 in § 7 that we need only prove the boundedness for all n of

$$\mathbf{M} \mid A_n(\omega)\, x \mid^2.$$

Let R_n denote a nonnegative symmetric operator such that

$$\mathbf{M} \mid A_n(\omega)\, x - x \mid^2 = (R_n x, x).$$

Then

$$\begin{aligned}
&(R_{n+1} x, x) = \mathbf{M} \mid A_{n+1}(\omega)\, x - x \mid^2 = \mathbf{M} \mid A_n(\omega)\, x + \\
&+ A_n(\omega) B_{n+1}(\omega)\, x - x \mid^2 = \mathbf{M} \mid A_n(\omega)\, x - x \mid^2 + \\
&+ 2\mathbf{M} (A_n(\omega)\, x - x,\ A_n(\omega) B_{n+1}(\omega)\, x) + \\
&+ \mathbf{M} \mid A_n(\omega) B_{n+1}(\omega)\, x \mid^2 = (R_n x, x) + \\
&+ \mathbf{M} (R_n B_{n+1}(\omega)\, x,\ B_{n+1}(\omega)\, x) + \mathbf{M} \mid B_{n+1}(\omega)\, x \mid^2,
\end{aligned}$$

since, by what was proved before,

$$\mathbf{M}\, (A_n(\omega)\, x - x,\ A_n(\omega) B_{n+1}(\omega)\, x) = 0,$$

and $\mathbf{M} \mid A_n(\omega) B_{n+1}(\omega)\, x \mid^2 = \mathbf{M}\mathbf{M} (\mid A_n(\omega) B_{n+1}(\omega)\, x \mid^2 / \mathfrak{F}_{B_{n+1}}) =$

$$= \mathbf{M} \{\mathbf{M} \mid A_n(\omega)\, z \mid^2_{z = B_{n+1}(\omega) x}\} = \mathbf{M} \{[\mid z \mid^2 + (R_n z, z)]_{z = B_{n+1}(\omega) x}\}.$$

Let $\| R_n \|$ be the norm of R_n. Then, since $0 \leqslant R_n \leqslant R_{n+1}$, it follows that $\| R_{n+1} \| \geqslant \| R_n \|$ and

$$\mathbf{M} (R_n B_{n+1}(\omega)\, x,\ B_{n+1}(\omega)\, x) \leqslant \| R_n \| \mathbf{M} \mid B_{n+1}(\omega)\, x \mid^2 =$$

$$= \| R_n \| (V_{n+1} x, x).$$

Thus,

$$(R_{n+1}x, x) \leqslant (R_n x, x) + (1 + \|R_n\|)(V_{n+1}x, x),$$

and for $n > l$

$$(R_{n+1}x, x) \leqslant (R_l x, x) + (1 + \|R_n\|) \sum_{k=l+1}^{n+1} (V_k x, x).$$

Hence

$$\|R_{n+1}\| \leqslant \|R_l\| + (1 + \|R_n\|) \| \sum_{k=l+1}^{n+1} V_k \| \leqslant \|R_l\| +$$

$$+ (1 + \|R_n\|) \| \sum_{k=l+1}^{\infty} V_k \| \leqslant \|R_l\| + (1 + \|R_{n+1}\|) \| \sum_{k=l+1}^{\infty} V_k \|.$$

Therefore,

$$\|R_{n+1}\| \leqslant \left(1 - \left\| \sum_{k=l+1}^{\infty} V_k \right\|\right)^{-1} \left(\|R_l\| + \left\| \sum_{k=l+1}^{\infty} V_k \right\|\right). \quad ■$$

We now consider the infinite product with the order of the factors changed:

$$\prod_{k=\infty}^{1} (I + B_k(\omega)) = \lim_{n\to\infty} \prod_{k=n}^{1} (I + B_k(\omega)) =$$

$$= \lim_{n\to\infty} (I + B_n(\omega)) \ldots (I + B_1(\omega)). \qquad (3.22)$$

Note that this infinite product is more natural than the preceding one, since upon application to an element x the operators $(I + B_k(\omega))$ act successively in increasing order of their indices. If $B_k(\omega) \in \mathbf{L}_s(\Omega, \mathbf{X})$, then also $\prod_{k=n}^{1} (I + B_k(\omega)) \in \mathbf{L}_s(\Omega, \mathbf{X})$. If $\mathbf{M}B_k(\omega)x = 0 \; \forall x \in \mathbf{X}$, then the sequence

$$\tilde{A}_n(\omega) = \prod_{k=n}^{1} (I + B_k(\omega)) \qquad (3.23)$$

is a strong operator-valued martingale. Indeed,

$$\tilde{A}_{n+1}(\omega) = \tilde{A}_n(\omega) + B_{n+1}(\omega)\tilde{A}_n(\omega), \qquad (3.24)$$

and

$$\mathbf{M}\,[(B_{n+1}(\omega)\,\tilde{A}_n(\omega)\,x,\,y)/\mathfrak{f}_n] = \mathbf{M}\,(B_{n+1}(\omega)\,z,\,y)\,\big|_{z=\tilde{A}_n(\omega)x} = 0.$$

If the operators $B_n(\omega)$ are such that $B_n^*(\omega) \in \mathbf{L}_s(\Omega, \mathbf{X})$, then

$$\tilde{A}_n^*(\omega) = \prod_{k=1}^{n} (I + B_k^*(\omega)),$$

and the convergence of this infinite product can be investigated with the help of Theorem 1 (note, however, that strong convergence of $\tilde{A}_n^*(\omega)$ implies only weak convergence of $\tilde{A}_n(\omega)$). To prove that the sequence $\tilde{A}_n(\omega)$ is strongly convergent, we need conditions more stringent than those required in Theorem 1.

Theorem 2. Let $B_n(\omega)$ be independent, $(V_n x, x) = \mathbf{M}\,|B_n(\omega)x|^2$. If

$$\sum_{n=1}^{\infty} \|V_n\| < \infty,$$

then the sequence $\tilde{A}_n(\omega)x$, defined by (3.23) has a limit with probability 1, whatever $x \in \mathbf{X}$.

Proof. As in Theorem 1, it will suffice to prove that $\mathbf{M}\,|\tilde{A}_{n+1}(\omega)x|^2$ is bounded. From (3.23) we obtain

$$\mathbf{M}\,|\tilde{A}_{n+1}(\omega)\,x|^2 = \mathbf{M}\,|\tilde{A}_n(\omega)\,x|^2 + \mathbf{M}\,|B_{n+1}(\omega)\,\tilde{A}_n(\omega)|^2 =$$
$$= \mathbf{M}\,|\tilde{A}_n(\omega)\,x|^2 + \mathbf{M}\,(V_{n+1}\tilde{A}_n(\omega)\,x,\ \tilde{A}_n(\omega)\,x) \leqslant$$
$$\leqslant \mathbf{M}\,|\tilde{A}_n(\omega)\,x|^2 (1 + \|V_{n+1}\|).$$

Consequently,

$$\mathbf{M}\,|\tilde{A}_{n+1}(\omega)\,x|^2 \leqslant \left(\prod_{k=1}^{n+1} (1 + \|V_k\|)\,|x|^2 \leqslant \prod_{k=1}^{\infty} (1 + \|V_k\|)\,|x|^2 \leqslant\right.$$
$$\leqslant \exp\left\{\sum_{k=1}^{\infty} \|V_k\|\right\} |x|.$$

8.2. Convergence of Infinite Products given the Existence of Two Moments

Consider the infinite product (3.18) with independent $B_k(\omega) \in \mathbf{L}_s(\Omega, \mathbf{X})$, such that $\mathbf{M} | B_k(\omega)x|^2 < \infty, \forall x \in \mathbf{X}$. Then there exist bounded operators U_k and V_k such that

$$\mathbf{M}(B_k(\omega)x, y) = (U_k x, x), \quad \mathbf{M} | B_k(\omega)x - U_k x|^2 = (V_k x, x),$$

with V_k symmetric and nonnegative. (In contrast to the previous section, it is not assumed that $U_k = 0$.)

We wish to investigate the convergence of the infinite product

$$\prod_{k=\infty}^{1} (I + B_k(\omega)) = \lim_{n\to\infty} (I + B_n(\omega))(I + B_{n-1}(\omega)) \ldots$$
$$\ldots (I + B_1(\omega)). \tag{3.25}$$

Theorem 3. Assume the following conditions satisfied:

(1) for some l, the infinite product of non-random operators

$$\prod_{k=\infty}^{l} (I + U_k)$$

is convergent in operator norm to a nondegenerate operator.

(2) $\sum_{k=1}^{\infty} \|V_k\| < \infty$.

Then the infinite product (3.25) is strongly convergent with probability 1; that is to say, for all $x \in \mathbf{X}$ the following limit exists with probability 1:

$$\lim_{n\to\infty} (I + B_n(\omega))(I + B_{n-1}(\omega)) \ldots (I + B_1(\omega))x. \tag{3.26}$$

Proof. That the expression after the limit symbol in (3.26) is defined follows from the results of the preceding section. Obviously, it will suffice to prove the existence of the limit

$$\lim_{n\to\infty} (I + B_n(\omega))(I + B_{n-1}(\omega)) \ldots (I + B_l(\omega))x,$$

where l is the number whose existence is demanded in condition (1). We may therefore assume that condition (1) holds for $l = 1$. Then all the operators $(I + U_k)$ are nondegenerate. Put

$$A_n(\omega) = (I + U_1)^{-1} \dots (I + U_n)^{-1} (I + B_n(\omega)) \dots (I + B_1(\omega)).$$

Then

$$A_{n+1}(\omega) = (I + U_1)^{-1} \dots (I + U_{n+1})^{-1} (I + B_{n+1}(\omega)) \times$$
$$\times (I + U_n) \dots (I + U_1) A_n(\omega) = (I + \hat{B}_{n+1}(\omega)) A_n(\omega),$$

where

$$\hat{B}_{n+1}(\omega) = (I + U_1)^{-1} \dots (I + U_{n+1})^{-1} (B_{n+1}(\omega) -$$
$$- U_{n+1})(I + U_n) \dots (I + U_1).$$

Clearly, $\hat{B}_{n+1}(\omega) \in \mathbf{L}_s(\Omega, \mathbf{X})$. In addition,

$$\mathbf{M}(\hat{B}_{n+1}(\omega) x, y) = \mathbf{M}([B_{n+1}(\omega) - U_{n+1}](I + U_n) \dots$$
$$\dots (I + U_1) x, [(I + U_1)^{-1} \dots (I + U_{n+1})^{-1}]^* y) = 0,$$
$$\mathbf{M} |\hat{B}_n(\omega) x|^2 \leqslant \|(I + U_1)^{-1} \dots (I + U_n)^{-1}\|^2 \mathbf{M} |B_n(\omega)(I +$$
$$+ U_{n-1}) \dots (I + U_1) x|^2 \leqslant \|(I + U_1)^{-1} \dots$$
$$\dots (I + U_n)^{-1}\|^2 \|(I + U_{n-1}) \dots (I + U_1)\|^2 \|V_n\|.$$

Condition (1) of our theorem implies the boundedness of the expressions

$$\|(I + U_n) \dots (I + U_1)\| \text{ and } \|(I + U_n)^{-1} \dots (I + U_1)^{-1}\|.$$

Therefore, for some $\gamma > 0$,

$$\mathbf{M} |\hat{B}_n(\omega) x|^2 < \gamma \|V_n\| |x|^2.$$

Let $\hat{V}_n$ be a nonnegative symmetric operator such that

$$\mathbf{M} |\hat{B}_n(\omega) x|^2 = (\hat{V}_n x, x).$$

Then $\|\hat{V}_n\| \leqslant \gamma \|V_n\|$. Since

$$A_n(\omega) = (I + \hat{B}_n(\omega)) \ \ldots \ (I + \hat{B}_1(\omega)),$$

it follows from Theorem 2 that $A_n(\omega)\, x$ has a limit with probability 1. Note that

$$(I + B_n(\omega)) \ \ldots \ (I + B_1(\omega))\, x = \left(\prod_{k=n}^{1} (I + \dot{U}_k)\right) A_n(\omega)\, x.$$

Using condition (1), we complete the proof of the theorem.■

8.3. Convergence of Infinite Products in Absolute Norm

Let us assume that the operators $B_n(\omega)$ are such that

$$\| B_n(\omega) \|_2^2 = \operatorname{sp} B_n^*(\omega)\, B_n(\omega) < \infty \tag{3.27}$$

with probability 1, and investigate the convergence of infinite products of type (3.18) or (3.22) in the norm $\|\cdot\|_2$ defined by (3.27). This norm is known as the absolute operator norm (or trace norm). The case is of some interest as it enables necessary and sufficient conditions to be obtained for convergence.

Theorem 4. Let $B_n(\omega) \in \mathbf{L}(\Omega, \mathbf{X})$ be independent; assume that condition (3.27) holds for all n, and that the operator $I + B_n(\omega)$ is nondegenerate for all n. For $\delta > 0$, define

$$B_n^\delta(\omega) = \begin{cases} B_n(\omega), & \text{if } \operatorname{sp} B_n^*(\omega)\, B_n(\omega) \leqslant \delta; \\ 0, & \text{if } \operatorname{sp} B_n^*(\omega)\, B_n(\omega) > \delta, \end{cases} \qquad U_n^\delta = \mathbf{M} B_n^\delta(\omega).$$

Then a necessary condition for the infinite product (3.22) to converge in $\mathbf{L}(\Omega, \mathbf{X})$ to a nondegenerate operator is that, for $\delta < 1$:

(1) the infinite product

$$\prod_{n=\infty}^{1} (I + U_n^\delta)$$

is convergent in $\|\cdot\|_2$ to a nondegenerate operator;
(2) the series $\Sigma \mathbf{M} \operatorname{sp} (B_n^\delta(\omega) - U_n^\delta)^* (B_n^\delta(\omega) - U_n^\delta)$; is convergent;
(3) $\Sigma \mathbf{P} \{\operatorname{sp} B_n^*(\omega)\, B_n(\omega) > \delta\} < \infty$;
a sufficient condition for the above assertion to be true is that these conditions hold for some $\delta > 0$.

Proof. Necessity. It is easy to check that if the product of the non-random operators $\prod_{k=\infty}^{1}(I+A_k)$ is convergent in $\|\cdot\|_2$, then $\|A_k\|_2\to 0$. Hence it follows from the fact that the infinite product (3.22) is convergent in $\|\cdot\|_2$ with probability 1 that $\operatorname{sp} B_n^*(\omega)B_n(\omega)\to 0$ with probability 1; as the r.v.'s $\operatorname{sp} B_n^*(\omega)B(\omega)$ are independent, we have, $\forall\delta>0$,

$$\sum_n \mathbf{P}\{\operatorname{sp} B_n^*(\omega)B_n(\omega)>\delta\}<\infty$$

(see, e.g., Theorem 5 in [8, p. 91]). This proves the necessity of condition (3). Conversely, if it holds, then the product (3.22) differs from

$$\prod_{\infty}^{1}(I+B_n^{\delta}(\omega)) \qquad (3.28)$$

in at most a finite number of (nondegenerate) factors (from some n on, $B_n^{\delta}(\omega)=B_n(\omega)$), and so this product is uniformly convergent. As the product (3.28) is convergent, it follows that

$$\eta_n=\left\|\prod_{k=n}^{1}(I+B_k^{\delta}(\omega))-I\right\|_2$$

are bounded in probability. We shall show that then these r.v.'s have uniformly bounded moments. Define

$$\eta_{l,n}=\left\|\prod_{k=n}^{l+1}(I+B_k^{\delta}(\omega))-I\right\|_2,\quad l<n.$$

It is readily seen that for all $l<n$

$$\eta_n\leqslant(1+\eta_l)(1+\eta_{l,n})-1.$$

Let ν be a number such that η_ν is the first element of the sequence greater than $\alpha>0$. Then, since $\eta_{\nu-1}<\alpha$, $\eta_{\nu-1,\nu}<\delta$, it follows that $\eta_\nu<(1+\alpha)(1+\delta)-1=\alpha+\delta(1+\alpha)$. Now, if ν_1 is such that η_{ν_1} is the first element greater than $\beta\,(\beta>(1+\alpha)(1+\delta))$, then $\nu<\nu_1$, and so

$$(1+\alpha)(1+\delta)(1+\eta_{\nu,\nu_1})>\beta;$$

$$\eta_{\nu,\nu_1}>\frac{\beta}{(1+\alpha)(1+\delta)}-1.$$

Thus,

$$\mathbf{P}\{\sup_n \eta_n > \beta - 1\} \leqslant$$

$$\leqslant \mathbf{P}\{\sup_n \eta_n > \alpha\}\, \mathbf{P}\left\{\sup_{n,m} \eta_{n,m} > \frac{\beta}{(1+\alpha)(1+\delta)}\right\}.$$

Setting $\beta = (1+\alpha)^2(1+\delta)$, we find that

$$\mathbf{P}\{\sup_n \eta_n > (1+\alpha)^2(1+\delta)\} \leqslant \mathbf{P}^2\{\sup_{m>n>0} \eta_{n,m} > \alpha\}.$$

Proceeding in similar fashion, we show that

$$\mathbf{P}\{\sup_n \eta_n > (1+\alpha)^k(1+\delta)^{k-1} - 1\} \leqslant \mathbf{P}^k\{\sup_{m>n\geqslant 0} \eta_{n,m} > \alpha\}.$$

Similarly,

$$\mathbf{P}\{\sup_{n\geqslant l} \eta_{l,n} > (1+\alpha)^k(1+\delta)^{k-1} - 1\} \leqslant \mathbf{P}^k\{\sup_{m>n\geqslant l} \eta_{n,m} > \alpha\}.$$

The fact that the product (3.28) is convergent in $\|\cdot\|_2$ implies that, for any $\alpha > 0$, we can find l such that, if $m > n \geqslant l$

$$\mathbf{P}\{\sup_{m>n\geqslant l} \eta_{m,n} > \alpha\} < \varepsilon.$$

Consequently, if ε is so chosen that

$$\Sigma\,[(1+\alpha)^k(1+\delta)^{k-1}]^r\,\varepsilon^k < \infty,$$

then

$$\mathbf{M}(\sup_{n\geqslant l} \eta_{l,n})^r \leqslant \sum_{k=1}^{\infty} [(1+\alpha)^k(1+\delta)^{k-1}]^r\, \mathbf{P}\{\sup_{n\geqslant l} \eta_{l,n} >$$

$$> (1+\alpha)^{k-1}(1+\delta)^{k-2} - 1\} \leqslant \sum_{k=1}^{\infty} [(1+\alpha)^k\; 1+\delta)^{k-1}]^r\,\varepsilon^k.$$

At the same time, $\eta_l \leqslant (1+\delta) - 1$. Consequently,

$$\mathbf{M}|\sup_n \eta_n|^r < \infty. \tag{3.29}$$

Next,

$$\left\| \prod_{k=n}^{1} (I + \mathbf{M}B_k^\delta(\omega)) - \prod_{k=m}^{1} (I + \mathbf{M}B_k^\delta(\omega)) \right\|_2 =$$

$$= \left\| \mathbf{M} \left(\prod_{k=n}^{1} (I + B_k^\delta(\omega)) \right) - \prod_{k=m}^{1} (I + B_k^\delta(\omega)) \right\|_2 \leqslant$$

$$\leqslant \mathbf{M} \left\| \prod_{k=n}^{1} (I + B_k^\delta(\omega)) - \prod_{k=m}^{1} (I + B_k^\delta(\omega)) \right\|_2 \to 0$$

as $m, n \to \infty$, since

$$\left\| \prod_{k=n}^{1} (I + B_k^\delta(\omega)) - \prod_{k=m}^{1} (I + B_k^\delta(\omega)) \right\|_2 \to 0$$

in probability, and moreover this r.v. is majorized by the expression

$$2 \sup_n \left\| \prod_{k=n}^{1} (I + B_k^\delta(\omega)) - I \right\| = 2 \sup_n \eta_n,$$

and $\mathbf{M} \sup_n \eta_n < \infty$, as shown above. Thus,

$$\lim_{n,m\to\infty} \left\| \prod_{k=n}^{1} (I + \mathbf{M}B_k^\delta(\omega)) - \prod_{k=m}^{1} (I + \mathbf{M}B_k^\delta(\omega)) \right\|_2 = 0.$$

This proves necessity for condition (1). It follows from (3.29) that

$$\rho_n = \mathbf{M}\eta_n^2 = \mathbf{M}\,\mathrm{sp}\left(\prod_{k=n}^{1} (I + B_k^\delta(\omega)) - I \right)^* \left(\prod_{k=n}^{1} (I + B_k^\delta(\omega)) - I \right) =$$

$$= \mathbf{M} \left\| \prod_{k=n}^{1} (I + B_k^\delta(\omega)) - I \right\|_2^2$$

is bounded. But then the same is true of

$$\mathbf{M} \left\| \left(\prod_{k=n}^{1} (I + U_k^\delta) \right)^{-1} \prod_{k=n}^{1} (I + B_k^\delta(\omega)) - I \right\|_2^2,$$

since

$$\left\| \left(\prod_{k=n}^{1} (I + U_k^\delta) \right)^{-1} \prod_{k=n}^{1} (I + B_k^\delta(\omega)) - I \right\|_2 \leqslant$$

$$\leqslant \left\| \left(\prod_{k=n}^{1} (I + U_k^\delta) \right)^{-1} \right\| \left\| \prod_{k=n}^{1} (I + B_k^\delta(\omega)) - I \right\|_2 +$$

$$+\left\|\left(\prod_{k=n}^{1}(I+U_k^{\delta})\right)^{-1}-I\right\|_2,$$

where the last term on the right is bounded since $\prod_{k=n}^{1}(I+U_k^{\delta})$ is convergent in $\|\cdot\|_2$ to a nondegenerate operator, and so $(\prod_{k=n}^{1}(I+U_k^{\delta}))^{-1}$ is convergent in $\|\cdot\|_2$ to the inverse of $\prod_{k=\infty}^{1}(I+U_k^{\delta})$. Moreover,

$$\left\|\left(\prod_{k=n}^{1}(I+U_k^{\delta})\right)^{-1}\right\|\leqslant 1+\left\|\left(\prod_{k=n}^{1}(I+U_k^{\delta})\right)^{-1}-I\right\|_2.$$

If we put

$$\left(\prod_{k=n}^{1}(I+U_k^{\delta})\right)^{-1}\prod_{k=n}^{1}(I+B_k^{\delta}(\omega))-I=C_n(\omega),$$

then (see § 8, subsection 2)

$$(I+C_{n+1}(\omega))=(I+\tilde{B}_{n+1}^{\delta}(\omega))(I+C_n(\omega)),$$

where (3.30)

$$\tilde{B}_{n+1}^{\delta}(\omega)=\left[\prod_{k=n+1}^{1}(I+U_k^{\delta})\right]^{-1}(B_{n+1}^{\delta}(\omega)-U_{n+1}^{\delta})\prod_{k=n}^{1}(I+U_k^{\delta}).$$

Obviously, $\mathbf{M}\tilde{B}_{n+1}^{\delta}(\omega)=0$. Since $C_n(\omega)$ is a martingale, it follows that also $\mathbf{M}C_n(\omega)=\mathbf{M}C_0(\omega)=0$. As $C_n(\omega)$ and $\tilde{B}_{n+1}^{\delta}(\omega)$ are independent,

$$\mathbf{M}\,\mathrm{sp}\,C_n^*(\omega)\tilde{B}_{n+1}^{\delta}(\omega)=\mathbf{M}\,\mathrm{sp}\tilde{B}_{n+1}^{\delta*}(\omega)C_n(\omega)=$$
$$=\mathbf{M}\,\mathrm{sp}\,C_n^*(\omega)\tilde{B}_{n+1}^{\delta}(\omega)C_n(\omega)=\mathbf{M}\,\mathrm{sp}\,C_n^*(\omega)B_{n+1}^{\delta*}(\omega)C_n(\omega)=$$
$$=\mathbf{M}\,\mathrm{sp}\,B_{n+1}^{\delta*}(\omega)B_{n+1}^{\delta}(\omega)C_n(\omega)=\mathbf{M}C_n^*(\omega)B_{n+1}^{\delta*}(\omega)B_{n+1}^{\delta}(\omega)=0.$$

Hence

$$\mathbf{M}\,\mathrm{sp}\,C_{n+1}^*(\omega)C_{n+1}(\omega)=\mathbf{M}\,\mathrm{sp}\,(C_n(\omega)+\tilde{B}_{n+1}^{\delta}(\omega)+$$
$$+B_{n+1}^{\delta}(\omega)C_n(\omega))^*(C_n(\omega)+\tilde{B}_{n+1}^{\delta}(\omega)+\tilde{B}_{n+1}^{\delta}(\omega)C_n(\omega))=$$
$$=\mathbf{M}\,\mathrm{sp}\,[C_n^*(\omega)C_n(\omega)+\tilde{B}_{n+1}^{\delta*}(\omega)\tilde{B}_{n+1}^{\delta}(\omega)+$$

$$+ C_n^*(\omega)\tilde{B}_{n+1}^{\delta*}(\omega)\tilde{B}_{n+1}^{\delta}(\omega)C_n(\omega)] \geqslant$$

$$\geqslant \mathbf{M}\,\mathrm{sp}\,C_n^*(\omega)C_n(\omega) + \mathbf{M}\,\mathrm{sp}\,\tilde{B}_{n+1}^{\delta*}(\omega)\tilde{B}_{n+1}^{\delta}(\omega).$$

Thus,

$$\sum_{k=1}^{n} \mathbf{M}\,\mathrm{sp}\,\tilde{B}_{n+1}^{\delta*}(\omega)B_{n+1}^{\delta}(\omega) \leqslant \mathbf{M}\,\mathrm{sp}\,C_n^*(\omega)C_n(\omega),$$

and the expression on the left is bounded. We note that

$$\mathrm{sp}\,(UBV)^*(UBV) = \mathrm{sp}\,(V^*B^*U^*UB^*V) \leqslant$$

$$\leqslant \|V\|^2\,\mathrm{sp}\,B^*U^*UB = \|V\|^2\,\mathrm{sp}\,UBB^*U^* \leqslant \|V\|^2\|U\|^2\,\mathrm{sp}\,B^*B.$$

Therefore, in view of (3.30), we obtain

$$\mathrm{sp}\,(B_{n+1}^{\delta}(\omega) - U_{m+1}^{\delta})^*(B_{n+1}^{\delta}(\omega) - U_{n+1}^{\delta}) \leqslant$$

$$\leqslant \left\| \prod_{k=n+1}^{1} (I + U_k^{\delta}) \right\|^2 \left\| \left(\prod_{k=n}^{1} (I + U_k^{\delta}) \right)^{-1} \right\|^2 \mathrm{sp}\,\tilde{B}_{n+1}^{\delta}\tilde{B}_n^{\delta}.$$

Since the coefficients of $\mathrm{sp}\,\tilde{B}_{n+1}^{\delta*}(\omega)\tilde{B}_{n+1}^{\delta}(\omega)$ on the right are bounded (condition 1 has already been established), there exists a constant α such that

$$\sum_{k=1}^{\infty} \mathbf{M}\,\mathrm{sp}\,(B_k^{\delta}(\omega) - U_k^{\delta})^*(B_k^{\delta}(\omega) - U_k^{\delta}) \leqslant \alpha \sum_{k=1}^{\infty} \mathbf{M}\,\mathrm{sp}\,\tilde{B}_k^{\delta*}(\omega)\tilde{B}_k^{\delta}(\omega).$$

This proves necessity of condition 2.

Sufficiency. As remarked, when condition 3 is valid the infinite products

$$\prod_{\infty}^{1}(I + B_k(\omega)) \qquad \prod_{\infty}^{1}(I + B_k^{\delta}(\omega))$$

are either both convergent or both divergent. We shall prove that the second product is convergent. It follows from condition 1 that this will be true if we can prove the existence of a limit in $\|\cdot\|_2$ for the expression

$$\lim_{n\to\infty} \left(\prod_{k=n}^{1} (I + U_k^{\delta}) \right)^{-1} \prod_{k=n}^{1} (I + B_k^{\delta}(\omega)) = \prod_{\infty}^{1}(I + \tilde{B}_k^{\delta}(\omega)),$$

where $\tilde{B}_k^\delta(\omega)$ is expressed in terms of $B_k^\delta(\omega)$ by formula (3.30). It follows from that formula and from condition 2 that

$$\sum \mathbf{M}\,\mathrm{sp}\,\tilde{B}_k^{\delta}(\omega)\,\tilde{B}_k^{\delta}(\omega) < \infty. \qquad (3.31)$$

If $C_n(\omega) = \prod_n^1 (I + \tilde{B}_k^\delta(\omega)) - I,\ C_{n+1}(\omega) =$

$$= C_n(\omega) + \tilde{B}_{n+1}^\delta(\omega) + \tilde{B}_{n+1}^\delta(\omega)\,C_n(\omega),$$

then, as shown in the necessity proof,

$$\mathbf{M}\,\mathrm{sp}\,C_{n+1}^*(\omega)\,C_{n+1}^*(\omega) = \mathbf{M}\,\mathrm{sp}\,C_n^*(\omega)\,C_n^*(\omega) + \qquad .(3.32)$$

$$+ \mathbf{M}\,\mathrm{sp}\,\tilde{B}_{n+1}^{\delta*}(\omega)\,\tilde{B}_{n+1}^{\delta}(\omega) + \mathbf{M}\,\mathrm{sp}\,C_n^*(\omega)\,\tilde{B}_{n+1}^{\delta*}(\omega)\,\tilde{B}_{n+1}^{\delta}(\omega)\,C_n(\omega).$$

But

$$\mathbf{M}\,\mathrm{sp}\,C_n^*(\omega)\,\tilde{B}_{n+1}^{\delta}(\omega)\,\tilde{B}_{n+1}^{\delta}(\omega)\,C_n(\omega) \leqslant$$

$$\leqslant \mathbf{M}\,\|C_n(\omega)\|^2\,\mathrm{sp}\,\tilde{B}_{n+1}^{\delta*}(\omega)\,\tilde{B}_{n+1}^{\delta}(\omega) \leqslant$$

$$\leqslant \mathbf{M}\,\mathrm{sp}\,C_n^*(\omega)\,C_n(\omega)\,\mathrm{sp}\,\tilde{B}_{n+1}^{\delta*}(\omega)\,\tilde{B}^{\delta}{}_{+1}(\omega) =$$

$$= \mathbf{M}\,\mathrm{sp}\,C_n^*(\omega)\,C_n(\omega)\,\mathbf{M}\,\mathrm{sp}\,\tilde{B}_{n+1}^{\delta*}(\omega)\,\tilde{B}_{n+1}^{\delta}(\omega)$$

(because $C_n(\omega)$ and $B_{n+1}^\delta(\omega)$ are independent). Therefore,

$$1 + \mathbf{M}\,\mathrm{sp}\,C_{n+1}^*(\omega)\,C_{n+1}(\omega) \leqslant$$

$$\leqslant (1 + \mathbf{M}\,\mathrm{sp}\,C_n^*(\omega)\,C_n(\omega))\,(1 + \mathbf{M}\,\mathrm{sp}\,\tilde{B}_{n+1}^{\delta*}(\omega)\,\tilde{B}_{n+1}^{\delta}(\omega));$$

$$\mathbf{M}\,\mathrm{sp}\,C_n^*(\omega)\,C_n(\omega) \leqslant \prod_{k=1}^{\infty} (1 + \mathbf{M}\,\mathrm{sp}\,\tilde{B}_k^{\delta*}(\omega)\,\tilde{B}_k^{\delta}(\omega)) - 1 \qquad (3.33)$$

(the right-hand side of the last inequality is bounded by virtue of (3.31)). Thus, $C_n(\omega)$ is a martingale in the Hilbert space $\mathbf{X}_2$ of Hilbert-Schmidt linear operators (i.e., operators A such that $\mathrm{sp}\,A^*A < \infty$) with scalar product $(A, B) = \mathrm{sp}\,B^*A$ and norm $\|A\|_2$. Since $\mathbf{M}\,\|C_n(\omega)\|_2^2$ is bounded, the convergent of $C_n(\omega)$ in the norm $\|\cdot\|_2$ follows from the remark to Theorem 3 of § 7.

The fact that the product (3.22) is nondegenerate can be

proved by considering the product $\prod_{n=\infty}^{m} (I + B_n(\omega))$ for sufficiently large m. The product $\prod_{n=\infty}^{m} (I + B_n^\delta(\omega))$ is nondegenerate if and only if $\prod_{n=\infty}^{m} (I + \tilde{B}_n^\delta(\omega))$ is nondegenerate, and the latter is nondegenerate if

$$\left\| \prod_{n=\infty}^{m} (I + \tilde{B}_n^\delta(\omega)) - I \right\| < 1.$$

Using (3.33) for the product $\prod_{n=\infty}^{m}$, we can write

$$\mathbf{P}\left\{ \left\| \prod_{n=\infty}^{m} (I + \tilde{B}_n^\delta(\omega)) - I \right\| < 1 \right\} \geqslant 1 -$$

$$- \mathbf{M}\,\mathrm{sp}\left(\prod_{n=\infty}^{m} (I + \tilde{B}_n^{\delta}(\omega)) - I \right)^* \left(\prod_{n=\infty}^{m} (I + \tilde{B}_n^\delta(\omega)) - I \right) \geqslant$$

$$\geqslant 2 - \prod_{n=m}^{\infty} (1 + \mathbf{M}\,\mathrm{sp}\,\tilde{B}_n^{\delta *}(\omega)\,\tilde{B}_n^\delta(\omega)).$$

Further,

$$\mathbf{P}\left\{ \prod_{n=\infty}^{m} (I + B_n(\omega)) \neq \prod_{n=\infty}^{m} (I + B_n^\delta(\omega)) \right\} \leqslant$$

$$\leqslant \sum_{n=m}^{\infty} \mathbf{P}\,\{\mathrm{sp}\,B_n^*(\omega)\,B_n(\omega) > \delta\}.$$

Consequently, the product $\prod_{n=\infty}^{m} (I + B_n(\omega))$ is nondegenerate with probability at least

$$2 - \prod_{n=m}^{\infty} (1 + \mathbf{M}\,\mathrm{sp}\,\tilde{B}_n^{\delta *}(\omega)\,\tilde{B}_n^\delta(\omega)) - \sum_{n=m}^{\infty} \mathbf{P}\,\{\mathrm{sp}\,B_n^*(\omega)\,B_n(\omega) > \delta\},$$

and this tends to 1 as $m \to \infty$. ■

Remark. If condition 1 in Theorem 4 is replaced by the following condition 1': the infinite product

$$\prod_{k=1}^{\infty} (I + U_k^\delta)$$

converges in B $\|\cdot\|_2$ to a nondegenerate operator, then the assertion of the theorem is valid for $\prod_{k=1}^{\infty}(I+B_k(\omega))$.

9. Continuous Operator-Valued Martingales

9.1. Some Properties of Continuous Real-valued Local Martingales

Let ξ_t, $t \geqslant 0$, be a continuous random process in $\mathbf{R}$. It is called a continuous local martingale relative to a filtration of σ algebras $\mathfrak{f}_t$ if there exists a sequence of Markov times τ_n (relative to the filtration $\mathfrak{f}_t$) such that $\tau_n \uparrow \infty$ with probability 1 and $\xi_{t\wedge\tau_n}$ is a martingale relative to $\mathfrak{f}_t$. If ξ_t is a continuous local martingale and

$$\zeta_n = \inf\{s : |\xi_s| > n\}, \ \inf \varnothing = +\infty,$$

then $\xi_{t\wedge\zeta_n}$ is also a martingale. This follows from the fact that $\xi_{t\wedge\tau_m\wedge\zeta_n}$ is a martingale for all m [9, p. 21], with

$$\mathbf{M}\,|\xi_{t\wedge\tau_m\wedge\zeta_n} - \xi_{t\wedge\tau_l\wedge\zeta_n}|^2 \to 0 \ \text{ as } \ m, l \to \infty,$$

since $\xi_{t\wedge\tau_m\wedge\zeta_n} - \xi_{t\wedge\tau_l\wedge\zeta_n} \to 0$ as $m \to \infty$ Therefore, $\xi_{t\wedge\zeta_n}$, as the limit in root square of martingales, is also a martingale. Thus, the definition of a continuous local martingale may be phrased equivalently with the times τ_n replaced by ζ_n.

For any continuous local martingale ξ_t, there exists a continuous nondecreasing process $\langle\xi\rangle_t$ ($\langle\xi\rangle_0 = 0$), such that $\xi_t^2 - \langle\xi\rangle_t$ is also a continuous local martingale (see [9, pp. 51-52; 56-57], Theorem 20). The process $\langle\xi\rangle_t$ is called the (quadratic) characteristic of the martingale ξ_t. If ξ_t and η_t are continuous local martingales, then the continuous process with bounded variation

$$\frac{1}{2}(\langle\xi+\eta\rangle_t - \langle\xi\rangle_t - \langle\eta\rangle_t) = \langle\xi, \eta\rangle_t \qquad (3.34)$$

has the property that $\xi_t\eta_t - \langle\xi, \eta\rangle_t$ is a local martingale; the process $\langle\xi, \eta\rangle_t$ is called the joint characteristic of ξ_t and η_t.

Lemma 1. The characteristic of a local martingale can be evaluated from the formula

$$\langle\xi\rangle_t = \lim_{n\to\infty} \sum_{k=0}^{n-1} \left(\xi_{\frac{k+1}{n}t} - \xi_{\frac{k}{n}t}\right)^2, \tag{3.35}$$

where the limit is to be understood in the sense of convergence in probability (see [9, p. 61], Theorem 22). It follows from formula (3.34) that the joint characteristic of continuous local martingales may be evaluated as a limit in probability

$$\langle\xi, \eta\rangle_t = \lim_{n\to\infty} \sum_{k=0}^{n-1} \left(\xi_{\frac{k+1}{n}t} - \xi_{\frac{k}{n}t}\right)\left(\eta_{\frac{k+1}{n}t} - \eta_{\frac{k}{n}t}\right). \tag{3.36}$$

If ξ_t is a continuous local martingale, then for all $\varepsilon > 0$ and $N > 0$,

$$\mathbf{P}\{\sup_{t\leqslant T} |\xi_t| > \varepsilon\} \leqslant \frac{N}{\varepsilon^2} + \mathbf{P}\{\langle\xi\rangle_T > N\} \tag{3.37}$$

(see [9, p. 63], Lemma 9). If ξ_t is a square-integrable martingale, i.e., $\sup_{t<\infty} \mathbf{M}\xi_t^2 < \infty$, then for any Markov time τ we have $\tau\,\mathbf{M}\,\xi_\tau^2 = \mathbf{M}\,\langle\xi\rangle_\tau$ ([9, p. 52], formula (44)). In the general case we can only say that

$$\mathbf{M}\xi_\tau^2 \leqslant \mathbf{M}\,\langle\xi\rangle_\tau.$$

However, if

$$\sup_t \mathbf{M}\xi_{t\wedge\tau}^2 < \infty, \quad \mathbf{M}\xi_\tau^2 = \mathbf{M}\,\langle\xi\rangle_\tau \tag{3.38}$$

(this follows from the fact that in this case $\xi_{t\wedge\tau}$ is a square-integrable martingale).

Theorem 1. Let $\xi_n(t)$ be a sequence of continuous local martingales such that $\langle\xi_n - \xi_m\rangle_t \to 0$ for all t as $m, n \to \infty$. Then there exists a continuous local martingale $\xi(t)$ such that $\langle\xi_n - \xi\rangle_t \to 0$ in probability. Moreover, for all $T > 0$,

$$\sup_{t\leqslant T} |\xi_n(t) - \xi(t)| \to 0, \quad \langle\xi_n\rangle_T \to \langle\xi\rangle_T.$$

Proof. Formula (3.35) implies the inequalities

$$\langle\xi_n\rangle_t \leqslant 2\,\langle\xi_m\rangle_t + 2\,\langle\xi_n - \xi_m\rangle_t;$$

$$|\langle\xi_n\rangle_t - \langle\xi_m\rangle_t| \leqslant \sqrt{\langle\xi_n - \xi_m\rangle_t\,(\langle\xi_n\rangle_t + \langle\xi_m\rangle_t)}. \tag{3.39}$$

It follows from the first inequality that $\langle\xi_n\rangle_t$ is bounded in probability, and from the second (using boundedness) that $\langle\xi_n\rangle_t$ converges to some nondecreasing process $\alpha(t)$. As the right-hand side is monotone in t, it follows that $\langle\xi_n\rangle_t$ converges uniformly to $\alpha(t)$ on every finite interval. Next, using formula (3.37), we obtain

$$\mathbf{P}\{\sup_{t\leqslant T}|\xi_n(t)-\xi_m(t)|>\varepsilon\}\leqslant\frac{\delta}{\varepsilon^2}+\mathbf{P}\{\langle\xi_n-\xi_m\rangle_T>\delta\}.$$

By assumption, $\mathbf{P}\{\langle\xi_n-\xi_m\rangle_T>\delta\}\to 0$ as $n, m\to\infty$ for all δ and T. Thus

$$\lim_{n,m\to\infty}\mathbf{P}\{\sup_{t\leqslant T}|\xi_n(t)-\xi_m(t)|>\varepsilon\}=0.$$

Thus, the sequence of processes $\xi_n(t)$ converges uniformly in probability to a continuous process $\xi(t)$. Choose a subsequence n_k such that

$$\mathbf{P}\left\{\sup_{t\leqslant k}|\xi_{n_k}(t)-\xi(t)|>\frac{1}{k}\right\}\leqslant\frac{1}{k^2};$$

$$\mathbf{P}\left\{\sup_{t\leqslant k}|\langle\xi_{n_k}\rangle_t-\alpha(t)|>\frac{1}{k}\right\}\leqslant\frac{1}{k^2}.$$

Then $\xi_{n_k}(t)\to\xi(t)$ uniformly on each finite interval, with probability 1. Consequently, the following expression is finite for all T with probability 1:

$$\eta_T=\sup_{t\leqslant T}\sup_k[|\xi_{n_k}(t)|,\ \langle\xi_{n_k}\rangle_t].$$

The sequence of Markov times

$$\tau_l=\inf\{T:\eta_T>l\}$$

has the property $\tau_l\to\infty$ as $l\to\infty$. Since $\xi_{n_k}(t\wedge\tau_l)$ are martingales and $\xi_{n_k}(t\wedge\tau_l)\to\xi(t\wedge\tau_l)$, while $\mathbf{M}|\xi_{n_k}(t\wedge\tau_l)|^2<l$, it follows that $\xi(t\wedge\tau_l)$ is also a martingale. Thus, $\xi(t)$ is a continuous local martingale, just as

$$\xi^2_{n_k}(t\wedge\tau_l)-\langle\xi_{n_k}\rangle_{t\wedge\tau_l}\to\xi^2(t\wedge\tau_l)-\alpha(t\wedge\tau_l)$$

and

$$\mathbf{M}|\xi^2_{n_k}(t\wedge\tau_l)-\langle\xi_{n_k}\rangle_{t\wedge\tau_l}|\leqslant l^2+l.$$

Thus $\alpha(t)$ is the characteristic of the continuous local martingale $\xi(t)$. ■

Remark. If $\xi_n(t)$ is a sequence of continuous local martingales such that for all T and $\varepsilon > 0$

$$\lim_{n\to\infty} \mathbf{P}\{\sup_{t\leq T} |\xi_n(t)| > \varepsilon\} = 0,$$

then $\langle\xi_n\rangle_t \to 0$ for all t. Indeed, if this were not true then, taking a suitable subsequence, we would have $\mathbf{P}\{\langle\xi_{n_k}\rangle_T > \delta\} > \delta$ for some $\delta > 0$.

Denote

$$\tau_k = \inf[s : |\xi_{n_k}(s)| > \rho].$$

Then, by (3.38),

$$\mathbf{M}\xi^2_{n_k}(\tau_k \wedge T) = \mathbf{M}\langle\xi_{n_k}\rangle_{T\wedge\tau_k}.$$

Thus

$$\rho^2 \geqslant \mathbf{M}\langle\xi_{n_k}\rangle_{T\wedge\tau_k} \geqslant \delta\mathbf{P}\{\{\langle\xi_{n_k}\rangle_T > \delta\} \cap \{\tau_k > T\}\} \geqslant$$

$$\geqslant \delta[\delta - \mathbf{P}\{\tau_k \leqslant T\}].$$

Now, for all $\rho > 0$; we have $\mathbf{P}\{\tau_k \leqslant T\} \to 0$ as $k \to \infty$; hence $\rho^2 \geqslant \delta^2$, whatever $\rho > 0$. This contradicts the fact that $\delta > 0$.

Thus, if a sequence of continuous local martingales $\xi_n(t)$ converges uniformly to some process on every finite interbal in probability, then $\xi_n(t) - \xi_m(t)$ converges to zero as $n, m \to \infty, \langle\xi_n - \xi_m\rangle_t \to 0$ and the assumptions of Theorem 1 are valid. Hence the limit process is a martingale, and $\langle\xi_n\rangle_t \to \langle\xi_0\rangle_t$ uniformly on any finite interval in probability.

9.2. Continuous Martingales with values in X

Consider an $\mathbf{X}$-valued continuous process $x(t, \omega)$, $t \geqslant 0$. It is called a continuous local martingale if, for any $y \in \mathbf{X}$ the process $(x(t, \omega), y)$ is a local martingale relative to a filtration of σ-algebras $\mathfrak{f}_t$, where $\mathfrak{f}_t$ is generated by $x(s, \omega)$, $s \leqslant t$. Denote the joint characteristic of the local martingales $(x(t, \omega), x)$ and $(x, (t, \omega), y)$ by $\alpha_t(\omega, x, y)$. It follows from (3.36) that

$$\alpha_t(\omega, x, y) = \lim_{n\to\infty} \sum_{k=0}^{n-1} \left(x\left(\frac{k+1}{n}t, \omega\right) - \right.$$

$$\left. - x\left(\frac{k}{n}t, \omega\right), x\right)\left(x\left(\frac{k+1}{n}t, \omega\right) - x\left(\frac{k}{n}t, \omega\right), y\right)$$

(limit in probability). Since the expression following the limit symbol is of the form $(A_t^{(n)}(\omega)\, x, y)$, where $A_t^{(n)}(\omega)$ is a random operator of the form

$$A_t^{(n)}(\omega) = \sum_{k=0}^{n-1} \left\langle x\left(\frac{k+1}{n}t, \omega\right) - \right.$$

$$\left. - x\left(\frac{k}{n}t, \omega\right) \circ x\left(\frac{k+1}{n}t, \omega\right) - x\left(\frac{k}{n}t, \omega\right)\right\rangle,$$

it follows from Theorem 1 of § 3 that

$$\alpha_t(\omega, x, y) = (A_t(\omega)\, x, y),$$

where $A_t(\omega) \in \mathbf{L}_w(\Omega, \mathbf{X})$. It is also clear that $(A_t(\omega)\, x, y) = (A_t(\omega)\, y, x), (A_t(\omega)\, x, x) \geqslant 0$, so that $A_t(\omega)$ is a nonnegative symmetric weak random operator. We denote this operator by $\langle x\rangle_t(\omega)$ and call it the operator characteristic of the local martingale $x(t, \omega)$. The following properties follow from the definition:

(1) $(\langle x\rangle_t(\omega)\, x, y)$ is continuous in t (with probability 1) for all $x, y \in \mathbf{X}$;

(2) $(\langle x\rangle_t(\omega)\, x, x)$ is a nondecreasing continuous function. We state a more important property as a theorem:

Theorem 2. Let $x(t, \omega)$ be a continuous local $\mathbf{X}$-valued martingale, $\langle x\rangle_t(\omega)$ its operator characteristic. If $\{e_k\}$ is an orthonormal basis, then

$$\sigma_t(\omega) = \sum_{k=1}^{\infty} (\langle x\rangle_t(\omega)\, e_k, e_k) \tag{3.40}$$

is defined and is a continuous nondecreasing function; under these conditions,

$$|x_t(\omega)|^2 - \sigma_t(\omega)$$

is also a local martingale.

Proof. We shall consider the following series, whose terms are continuous local real-valued martingales, and show that it converges to a martingale. We have

$$\sum_{k=1}^{\infty} [(x(t, \omega) e_k)^2 - (A_t(\omega) e_k, e_k)]$$

Now, for any n, denoting

$$\sum_{k=1}^{\infty} (x(t, \omega), e_k)^2 = |x(t, \omega)|^2 < \infty,$$

we obtain

$$\zeta_n = \inf [t : |x(t, \omega)| > n].$$

Hence, in view of the fact that $(A_{t\wedge\zeta_n}(\omega) e_k, e_k)$ is nonnegative, we see that the series

$$\sum_{k=1}^{\infty} (\langle x\rangle_{t\wedge\zeta_n}(\omega) e_k, e_k)$$

is convergent for all n, and hence the same is true of (3.40). Now, for $s < t$,

$$\sigma(s, \omega) - \sum_{k=1}^{m} (A_s(\omega) e_k, e_k) = \sum_{k=m+1}^{\infty} (A_s(\omega) e_k, e_k) \leqslant$$

$$\leqslant \sum_{k=m+1}^{\infty} (A_t(\omega) e_k, e_k) = \sigma(t, \omega) - \sum_{k=1}^{m} (A_t(\omega) e_k, e_k),$$

and so the series (3.40) is uniformly convergent; thus $\sigma(t, \omega)$ is a continuous nondecreasing function. Now let

$$\zeta_n' = \inf [t : \sigma(t, \omega) > n];$$

$$\tau_n = \zeta_n \wedge \zeta_n'.$$

Then $\tau_n \uparrow \infty$ as $n \to \infty$, with probability 1.

$$\sum_{k=1}^{m} [(x(t \wedge \tau_n, \omega), e_k)^2 - (\langle x\rangle_{t\wedge\tau_n}(\omega) e_k, e_k)]$$

is a sequence of martingales (the running index is m), whose terms are bounded by $2n$ (which is independent of m); this sequence converges to a process

$$|x(t \wedge \tau_n, \omega)|^2 - \alpha(t \wedge \tau_n, \omega),$$

which is also a martingale (for any n). Thus $|x(t, \omega)|^2 - \alpha(t, \omega)$ is a continuous local martingale.

Remark. Let $x(t, \omega)$ and $y(t, \omega)$ be continuous local martingales; then the same is true of the process $x(t, \omega) + y(t, \omega)$. Since

$$(x(t, \omega), y(t, \omega)) = \frac{1}{2}[|x(t, \omega) + y(t, \omega)|^2 - |x(t, \omega)|^2 - \\ - |y(t, \omega)|^2],$$

it follows that

$$(x(t, \omega), y(t, \omega)) - \frac{1}{2}\left(\sum_{k=1}^{\infty} \{(\langle x + y\rangle_t(\omega) e_k, e_k) - \\ - (\langle x\rangle_t(\omega) e_k, e_k) - (\langle y\rangle_t(\omega) e_k, e_k)\}\right)$$

is also a martingale. The expression

$$\langle x, y\rangle_t(\omega) = \frac{1}{2}\sum_{k=1}^{\infty} (\{\langle x + y\rangle_t(\omega) - \langle x\rangle_t(\omega) - \langle y\rangle_t(\omega)\} e_k, e_k)$$

is called the joint characteristic of continuous local $\mathbf{X}$-valued martingales. It may be evaluated as the limit in probability

$$\lim_{n\to\infty} \sum_{k=1}^{n} \left(x\left(\frac{k}{n}t, \omega\right) - x\left(\frac{k-1}{n}t, \omega\right), \\ y\left(\frac{k}{n}t, \omega\right) - y\left(\frac{k-1}{n}t, \omega\right)\right).$$

It will suffice to prove this for $y = x$. Let $\tau_\alpha = \inf[t; \langle x\rangle_t > \alpha]$. It will clearly suffice to show that

$$\langle x\rangle_{t\wedge\tau_\alpha} = \lim_{n\to\infty} \sum_{k=1}^{n} \left|x\left(\frac{k}{n}t \wedge \tau_\alpha, \omega\right) - x\left(\frac{k-1}{n}t \wedge \tau_\alpha, \omega\right)\right|^2 = \\ = \lim_{n\to\infty} \sum_{m=1}^{\infty} \sum_{k=1}^{n} \left(x\left(\frac{k}{n}t \wedge \tau_\alpha, \omega\right) - x\left(\frac{k-1}{n}t \wedge \tau_\alpha, \omega\right), e_m\right)^2,$$

and since

$$\mathbf{M}\sum_{k=1}^{n}\left(x\left(\frac{k}{n}t\wedge\tau_\alpha,\omega\right)-x\left(\frac{k-1}{n}t\wedge\tau_\alpha,\omega\right),e_m\right)^2=$$

$$=\mathbf{M}\left(\langle x\rangle_{t\wedge\tau_\alpha}(\omega)e_m,e_m\right),$$

$$\sum_m \mathbf{M}\left(\langle x\rangle_{t\wedge\tau_\alpha}(\omega)e_m,e_m\right)=\alpha<\infty,$$

we have

$$\lim_{n\to\infty}\sum_{m=1}^{\infty}\sum_{k=1}^{n}\left(x\left(\frac{k}{n}t\wedge\tau_\alpha,\omega\right)-x\left(\frac{k-1}{n}t\wedge\tau_\alpha,\omega\right),e_m\right)^2=$$

$$=\sum_{m=1}^{\infty}\lim_{n\to\infty}\sum_{k=1}^{n}\left(x\left(\frac{k}{n}t\wedge\tau_\alpha,\omega\right)-x\left(\frac{k-1}{n}t\wedge\tau_\alpha,\omega\right),e_m\right)^2=$$

$$=\sum_{m=1}^{\infty}\left(\langle x\rangle_{t\wedge\tau_\alpha}(\omega)e_m,e_m\right).$$

Theorem 3. Let $x_n(t,\omega)$ be a sequence of $\mathbf{X}$-valued continuous local martingales relative to a filtration $\mathfrak{f}_t$, $\langle x_n-x_m\rangle_t(\omega)$ the characteristic of the local martingale $x_n(t,\omega)-x_m(t,\omega)$. If there is a basis $\{e_k\}$ for which

$$\sum_{k=1}^{\infty}\left(\langle x_n-x_m\rangle_t(\omega)e_k,e_k\right) \tag{3.41}$$

is uniformly convergent in probability to zero on any finite interval, then there exists a continuous $\mathbf{X}$-valued local martingale $x_0(t,\omega)$ such that $x_n(t,\omega)\to x_0(t,\omega)$ for $x, y\in\mathbf{X}$, $(\langle x_n\rangle_t(\omega)x,y)\to(\langle x_0\rangle_t(\omega)x,y)$ and $\sum_{k=1}^{\infty}(\langle x_n\rangle(\omega)e_k,e_k)\to\sum_{k=1}^{\infty}(\langle x_0\rangle(\omega)e_k,e_k)$ uniformly in probability on any finite interval.

Proof. From inequalities (3.39) one readily shows that

$$\sum_{k=1}^{\infty}\left(\langle x_n\rangle_t(\omega)e_k,e_k\right)\ \text{is bounded relative to } n$$

in probability; consequently, the sequence

$$\sum_{k=1}^{\infty}\left(\langle x_n\rangle_t(\omega)e_k,e_k\right)$$

is uniformly convergent to some limit $\sigma(t,\omega)$ on every finite interval. Since

$$\mathbf{P}\{\sup_{t\leqslant T}|x_n(t,\omega)-x_m(t,\omega)|>\varepsilon\}\leqslant$$

$$\leqslant\frac{\delta}{\varepsilon^2}+\mathbf{P}\left\{\sum_{k=1}^{\infty}(\langle x_n-x_m\rangle_t(\omega)e_k,e_k)>\delta\right\},$$

it follows from the assumptions of the theorem that $x_n(t,\omega)$ converges uniformly on every finite interval to some limit $x_0(t,\omega)$. Moreover, it is then the case that $(x_n(t,\omega),x)$ converges for every $x\in\mathbf{X}$ to $(x_0(t,\omega),x)$; thus, by the remark to Theorem 1, the characteristic of $(x_n(t,\omega),x)$ also converges to $(x_0(t,\omega),x)$. Thus,

$$(\langle x_n\rangle_t(\omega)x,x)\to(\langle x_0\rangle_t(\omega)x,x),$$

$$(\langle x_n\rangle_t(\omega)x,y)=\frac{1}{2}[(\langle x_n\rangle_t(\omega)(x+y),x+y)-$$

$$-(\langle x_n\rangle_t(\omega)x,x)-(\langle x_n\rangle_t(\omega)y,y)]\to(\langle x_0\rangle_t(\omega)x,y)$$

uniformly on every finite interval in probability. Choosing a subsequence n_m such that $x_{n_m}(t,\omega)$ and $\sum_{k=1}^{\infty}(\langle x_{n_m}\rangle_t(\omega)e_k,e_k)$ converge uniformly with probability 1 on every finite interval, we show as in the proof of Theorem 1 that $x_0(t,\omega)$ is a martingale and

$$\sum_{k=1}^{\infty}(\langle x_0\rangle_t(\omega)e_k,e_k)=\sigma(t,\omega).\ \blacksquare$$

Remark. The following proposition is analogous to the remark following Theorem 1. Let $x_n(t,\omega)$ be a sequence of continuous $\mathbf{X}$-valued local martingales. If there exists an $\mathbf{X}$-valued process $x_0(t,\omega)$, such that for all $T>0$

$$\sup_{t\leqslant T}|x_n(t,\omega)-x_0(t,\omega)|$$

converges to zero in probability, then:

(a) $x_0(t,\omega)$ is a continuous local martingale;

(b) for all $t>0$ $\sum_{k=1}^{\infty}(\langle x_n-x_0\rangle_t(\omega)e_k,e_k)\to 0$ in probability;

(c) for all $x,y\in\mathbf{X}$ $(\langle x_n\rangle_t(\omega)x,y)\to(\langle x_0\rangle_t(\omega)x,y)$ uniformly in probability on every finite interval.

9.3. Operator-valued Continuous Martingales

Let $\{A_t(\omega),\ t \geqslant 0\}$ be a weak operator-valued martingale. We shall say that it is continuous if: (1) $A_t(\omega) \in \mathbf{L}_s(\Omega, \mathbf{X})$ for all $t \geqslant 0$;

(2) $A_t(\omega)x$ is a continuous process in $\mathbf{X}$ with probability 1.

If $A_t(\omega)x$ is considered as a function of the two variables t and x, then it is continuous in t and continuous in probability in x. It turns out that $\sup\limits_{t \leqslant T} |A_t(\omega)x - A_t(\omega)y| \to 0$ in probability for all $T > 0$ as $x - y \to 0$. This follows from the following proposition.

Lemma 2. If $A_t(\omega) \in \mathbf{L}_s(\Omega, \mathbf{X})$ for $t \geqslant 0$ and $A_t(\omega)x$ is continuous in x in probability, then for all $T > 0$

$$\psi(x) = \sup_{t \leqslant T} |A_t(\omega)x|$$

is bounded in probability when $|x| \leqslant 1$.

Proof. We put

$$\psi_n(x) = \sup_{k \leqslant n} |A_{\frac{kT}{n}}(\omega)x|.$$

The function $\psi_n(x)$ is bounded in probability as a function of x, and $\psi_n(x) \uparrow \psi(x)$. In addition,

$$\psi(x+y) \leqslant \psi(x) + \psi(y).$$

Using these properties, one proves the lemma along the same lines as the lemma of § 2. ■

It is readily seen that $A_t(\omega)x$ is a continuous local martingale in $\mathbf{X}$. Denote

$$\sigma_t(x, \omega) = \sum_{k=1}^{\infty} (\langle A(\omega)x\rangle_t e_k, e_k),$$

where $\langle A(\omega)x\rangle_t$ is the operator characteristic of the $\mathbf{X}$-valued local martingale $A_t(\omega)x$.

Consider the expression

$$\sigma_t(x, y, \omega) = \frac{1}{2}[\sigma_t(x+y, \omega) - \sigma_t(x, \omega) - \sigma_t(y, \omega)],$$

which has the following properties:

(1) $\sigma_t(x, y, \omega)$ is continuous in t, $\sigma_t(x, x, \omega)$ is a nondecreasing nonnegative function.

(2) for all $x; y \in \mathbf{X}, (A_t(\omega)x, A_t(\omega)y) - \sigma_t(x, y, \omega)$ is a real-valued local martingale relative to $\mathfrak{f}_t$. Indeed,

$$(A_t(\omega)x, A_t(\omega)y) - \sigma_t(x, y, \omega) = \frac{1}{2}[|A_t(\omega)(x+y)|^2 - \\ - \sigma_t(x+y, \omega) - (|A_t(\omega)x|^2 - \sigma_t(x, \omega)) - \\ - (|A_t(\omega)y|^2 - \sigma_t(y, \omega))],$$

and $|A_t(\omega)z|^2 - \sigma_t(z, \omega)$ is a local martingale relative to the filtration $\mathfrak{f}_t$, for all $z \in \mathbf{X}$

(3) If $x, y, z \in \mathbf{X}; \alpha, \beta \in \mathbf{R}$, then with probability 1

$$\sigma_t(\alpha x + \beta y, z, \omega) = \alpha\sigma_t(x, z, \omega) + \beta\sigma_t(y, z, \omega). \qquad (3.42)$$

Indeed,

$$(A_t(\omega)(\alpha x + \beta y), A_t(\omega)z) - \sigma_t(\alpha x + \beta y, z, \omega)$$

and

$$(A_t(\omega)(\alpha x + \beta y), A_t(\omega)z) - \alpha\sigma_t(x, z, \omega) - \beta\sigma_t(y, z, \omega)$$

are continuous local martingales. Thus the same holds for their difference

$$\sigma_t(\alpha x + \beta y, z) - \alpha\sigma_t(x, z, \omega) - \beta\sigma_t(y, z, \omega). \qquad (3.43)$$

As this difference is a continuous process of bounded variation, its characteristic (see (3.35)) is equal to zero. Hence it follows that the difference (3.43) vanishes. This proves (3.42), since $\sigma_0(x, y, \omega) = 0$.

(4) Using the symmetricity of $\sigma_t(x, y, \omega)$, one sees that

for $x_1, \ldots, x_k; y_1, \ldots, y_l \in \mathbf{X}, \alpha_1, \ldots, \alpha_k, \beta_1, \ldots, \beta_l \subset R$,

$$\sigma_t(\sum \alpha_i x_i, \sum \beta_j y_j, \omega) = \sum \alpha_i\beta_j\sigma_t(x_i, y_j, \omega).$$

<u>Theorem 4</u>. The function $\sigma_t(x, y, \omega)$ may be expressed as

$$\sigma_t(x, y, \omega) = (\langle A\rangle_t(\omega)x, y),$$

where $\langle A\rangle_t(\omega)$ is a random operator in $\mathbf{L}_w(\Omega, \mathbf{X})$ for any $t \geqslant 0$.

<u>Proof</u>. Put $A_t^{(n)}(\omega) = A_t(\omega)P_n$ where P_n is projection onto

the subspace spanned by $\{e_1, \ldots, e_n\}$, $\{e_k\}$ being some fixed basis in $\mathbf{X}$. Obviously, $A_t(\omega)P_n$ is also an operator-valued continuous martingale. If

$$\sigma_t^{(n)}(x, \omega) = \sum_{k=1}^{\infty} (\langle A^{(n)}(\omega)\, x\rangle_t e_k, e_k);$$

$$\sigma_t^{(n)}(x, y, \omega) = \frac{1}{2}[\sigma_t^{(n)}(x+y, \omega) - \sigma_t^{(n)}(x, \omega) - \sigma_t^{(n)}(y, \omega)],$$

then

$$\sigma_t^{(n)}(x, y, \omega) = \sigma_t(P_n x, P_n y, \omega).$$

Since $|x - P_n x| \to 0$ as $n \to \infty$, it follows from the lemma that for any $T > 0$

$$\sup_{t \leqslant T} |A_t(\omega)\, x - A_t^{(n)}(\omega)\, x| \to 0$$

in probability. Thus, by the remark to Theorem 3, $\sigma_t^{(n)}(x, \omega) \to \sigma_t(x, \omega)$, $\sigma_t^{(n)}(x, y, \omega) \to \sigma_t(x, y, \omega)$ in probability. Hence it will suffice to prove that

$$\sigma_t^{(n)}(x, y, \omega) = (\langle A^{(n)}\rangle_t(\omega)\, x, y), \tag{3.44}$$

where $\langle A^{(n)}\rangle_t(\omega) \in \mathbf{L}_w(\Omega, \mathbf{X})$, and then to use Theorem 1 of § 3. But it follows from property 4 above that

$$\sigma_t^{(n)}(x, y, \omega) = \sigma_t(P_n x, P_n y, \omega) =$$

$$= \sigma_t\left(\sum_{k=1}^{n} (x, e_k)\, e_k, \sum_{k=1}^{n} (y, e_k)\, e_k, \omega\right) =$$

$$= \sum_{k,j=1}^{n} \sigma_t(e_k, e_j, \omega)\, (x, e_k)\, (y, e_j).$$

Thus formula (3.44) is valid for $\sigma_t^{(n)}$ if one puts

$$\langle A^{(n)}\rangle_t(\omega) = \sum_{k,j=1}^{n} \sigma_t(e_k, e_j, \omega)\, \langle e_k \circ e_j\rangle.$$

It is clear that $\langle A^{(n)}\rangle_t \in \mathbf{L}(\Omega, \mathbf{X}) \subset \mathbf{L}_w(\Omega, \mathbf{X})$. ■

The operator-valued function $\langle A\rangle_t(\omega)$, which takes values in $\mathbf{L}_w^+(\Omega, \mathbf{X})$ (where $\mathbf{L}_w^+(\Omega, \mathbf{X})$ is the set of symmetric nonnegative weak random operators), is called the characteristic of the continuous operator-valued martingale $A_t(\omega)$. $\langle A\rangle_t(\omega)$ is nondecreasing as a function of t: for $t_1 < t_2$ we have

$\langle A\rangle_{t_2}(\omega) - \langle A\rangle_{t_1}(\omega) \in \mathbf{L}_w^+(\Omega, \mathbf{X})$.

As $A_t(\omega) \in \mathbf{L}_s(\Omega, \mathbf{X})$, this operator $A_t^*(\omega) A_t(\omega) \in \mathbf{L}_w^+(\Omega, \mathbf{X})$ is defined. It follows from property 2 that

$$A_t^*(\omega) A_t(\omega) - \langle A\rangle_t(\omega)$$

is a weak local martingale.

If $B_t(\omega) \in \mathbf{L}_w^+(\Omega, \mathbf{X}), B_{t_2}(\omega) - B_{t_1}(\omega) \in = \ni \mathbf{L}_w^+(\Omega, \mathbf{X})$, for $t_1 < t_2$, $B_0(\omega) = 0$, $B_t(\omega)$ is weakly continuous in t, and

$$A_t^*(\omega) A_t(\omega) - B_t(\omega)$$

is a weak local martingale, then $B_t(\omega) = \langle A\rangle_t(\omega)$. This follows from the fact that then $(B_t(\omega) - \langle A\rangle_t(\omega))$ is a continuous local martingale of bounded variation and so its characteristic vanishes.

Theorem 3 and the remark following it imply the following convergence theorem for continuous operator-valued martingales.

Theorem 5. Let $A_t^{(n)}(\omega)$ be a sequence of continuous operator-valued martingales relative to a filtration $\mathfrak{f}_t$.

1. If

$$\lim_{n,m\to\infty} \mathbf{P}\{\sup_{t\leqslant T}(\langle A^{(n)} - A^{(m)}\rangle_t(\omega) x, x) > \varepsilon\} = 0, \tag{3.45}$$

for all $x \in \mathbf{X}$, $T > 0$ and $\varepsilon > 0$, then there exists a continuous operator-valued martingale $A_t^{(0)}(\omega)$ relative to $\mathfrak{f}_t$, such that, for all $x \in \mathbf{X}$, $T > 0$ and $\varepsilon > 0$,

$$\lim_{n\to\infty} \mathbf{P}\{\sup_{t\leqslant T}|A_t^{(n)}(\omega) x - A_t^{(0)}(\omega) x| > \varepsilon\} = 0, \tag{3.46}$$

and moreover

$$\lim_{n\to} \mathbf{P}\{\sup_{t\leqslant T}|(\langle A^{(n)}\rangle_t(\omega) x, x) - (\langle A^{(0)}\rangle_t(\omega) x, x)| > \varepsilon\} = 0. \tag{3.47}$$

2. If $A_t^{(n)}(\omega)$ is a sequence for which (3.46) holds, then (3.45) and (3.47) are also valid for it.

9.4. Strong Operator-valued Wiener Processes

An operator-valued process $W_t(\omega)$, $t \geqslant 0$, with values in $\mathbf{L}_s(\Omega, \mathbf{X})$ is called a strong Wiener process if:

(a) $(W_t(\omega) x, y)$ is a Wiener process with mean 0 for all $x, y \in \mathbf{X}$ relative to the filtration $\mathfrak{f}_t$, generated by

$\{(W_s(\omega)u, z),\ s \leqslant t$ and $u, z \in \mathbf{X}\}$

(b) $W_t(\omega)x$ is a continuous process in $\mathbf{X}$ with probability 1.

Obviously, $W_t(\omega)$ is a continuous operator-valued martingale. Since the characteristic of the real-valued process $(W_t(\omega)x, y)$ coincides with

$$\mathbf{M}(W_t(\omega)x, y)^2 = t\mathbf{M}(W_1(\omega)x, y)^2,$$

it follows that the characteristic $\langle W\rangle_t(\omega)$ has the form

$$(\langle W\rangle_t(\omega)x, x) = t\sum_{k=1}^{\infty}\mathbf{M}(W_1(\omega)x, e_k)^2 = t\mathbf{M}\,|W_1(\omega)x|^2.$$

Thus,

$$\langle W\rangle_t(\omega) = tB,$$

where B is a bounded symmetric nonnegative operator,

$$B = \mathbf{M}W_1^*(\omega)W_1(\omega).$$

CHAPTER 4

STOCHASTIC INTEGRALS AND EQUATIONS

10. Stochastic Integral with Respect to an X-Valued Martingale

10.1. Definition

Let $x(t, \omega)$ be a continuous local martingale with values in $\mathbf{X}$, adapted to a filtration of σ-algebras $\mathfrak{f}_t$. Consider a function $z(t, \omega)$ with values in $\mathbf{X}$, defined and measurable on $\{[0, \infty) \times \Omega, \mathfrak{A}_+ \times \mathfrak{S}\}$, where $\mathfrak{A}_+$ is the σ-algebra of Borel subsets on $[0, \infty]$. Let us assume that $x(t, \omega)$ is $\mathfrak{f}_t$-measurable for all $t \geqslant 0$. We shall construct the stochastic integral

$$\int_0^T (z(t, \omega), dx(t, \omega)). \tag{4.1}$$

Let $\langle x \rangle_t(\omega)$ denote the characteristic of the martingale $x(t, \omega)$. Our definition of the integral (4.1) will follow the usual lines [7]. Let $\mathcal{F}$ be the set of all functions $z(t, \omega)$, with the above measurability properties, and $\mathcal{F}^0$ the subset of step functions in $\mathcal{F}$, i.e., functions $z(t, \omega)$ such that there is a sequence of points $0 = t_0 < t_1 < \dots$ $\dots < t_k < \dots$, with $z(t, \omega) = z(t_k, \omega)$, $t_k \leqslant t < t_{k+1}$. If $t_m < T \leqslant t_{m+1}$, we set

$$I(z, x)_T = \int_0^T (z(t, \omega), dx(t, \omega)) =$$

$$= \sum_{k=0}^{m-1} (z(t_k, \omega), x(t_{k+1}, \omega) - x(t_k, \omega)) +$$

$$+ (z(t_m, \omega), x(T, \omega) - x(t_m, \omega)). \tag{4.2}$$

The right-hand side is continuous as a function of T. Note the following properties of the function $I(z, x)_T$:

I. $I(z, x)_T$ is a continuous local martingale with respect to $\mathfrak{f}_t$.

II. If $z'(t, \omega) \in \mathcal{F}^0$, then the joint characteristic of the martingales $I(z, x)_t$ and $I(z', x)_t$ is given by the formula

$$\langle I(z, x), I(z', x)\rangle_t = \int_0^t (d\langle x\rangle_s(\omega) z(s, \omega), z'(s, \omega)), \quad (4.3)$$

where the integral on the right is defined for the step functions $z(s, \omega)$ and $z'(s, \omega)$ in $\mathcal{F}^0$ by the equality

$$\int_0 (d\langle x\rangle_s(\omega) z(s, \omega), z'(s, \omega)) =$$
$$= \sum_{k=0}^{m-1} ([\langle x\rangle_{t_{k+1}}(\omega) - \langle x\rangle_k(\omega)] z(t_k, \omega), z'(t_{k-1}, \omega)) +$$
$$+ ([\langle x\rangle_t(\omega) - \langle x\rangle_{t_m}(\omega)] z(t_m, \omega), z'(t_m, \omega)), \quad (4.4)$$

where $z(s, \omega)$ and $z'(s, \omega)$ are constant on the intervals $[t_0, t_1), [t_1, t_2,), \ldots, [t_m, t_{m+1})$ and $t \in (t_m, t_{m+1})$. The validity of this formula follows from the fact that the increment

$$\langle I(z, x), I(z', x)\rangle_t - \langle I(z, x), I(z', x)\rangle_{t_m}$$

on $[t_m, t_{m+1}]$ is identical to the joint characteristic of the martingales

$$(z(t_m, \omega), x(t, \omega) - x(t_m, \omega)); \quad (z'(t_m, \omega), x(t, \omega) - x(t_m, \omega))$$

which is equal to

$$([\langle x\rangle_t(\omega) - \langle x\rangle_{t_m}(\omega)] z(t_m, \omega), z'(t_m, \omega))$$

by virtue of the definition of $\langle x\rangle_t(\omega)$ and the $\mathfrak{f}_{t_m}$-measurability of $z(t_m, \omega)$ and $z'(t_m, \omega)$.

III. $I(z, x)_t$ is an additive and homogeneous functional in z; i.e., if $z, z' \in \mathcal{F}$, $\alpha, \alpha' \in \mathbf{R}$, then with probability 1

$$I(\alpha z + \alpha' z', x)_t = \alpha I(z, x)_t + \alpha' I(z', x)_t.$$

Let us assume that $z_n(t, \omega) \in \mathcal{F}^0$ is a sequence such that, for given $T > 0$ (in probability),

$$\lim_{n,m\to\infty} \int_0^T (d\langle x\rangle_t(\omega) [z_n(t, \omega) - z_m(t, \omega)], z_n(t, \omega) - z_m(t, \omega)) = 0. \quad (4.5)$$

Then, by Theorem 1 of § 9, we can state that there exists a continuous local martingale I_t adapted to the filtration $\mathfrak{f}_t$, such that

$$\sup_{t \leqslant T} | I_t - I(z_n, x)_t | \to 0$$

in probability. By virtue of the same theorem, the characteristic of this martingale is the limit in probability of the characteristic of $I(z_n, x)$, i.e.,

$$\langle I \rangle_t = \lim_{n \to \infty} \int_0^t (d\langle x \rangle_s(\omega) z_n(s, \omega),\ z_n(s, \omega)).$$

If a sequence $z_n(t, \omega)$ satisfying (4.5) can be assigned some limit $z(t, \omega) \in \mathcal{F}$, then it is natural to put

$$\int_0^T (z(t, \omega),\ dx(t, \omega)) = \lim_{n \to \infty} \int_0^T (z_n(t, \omega),\ dx(t, \omega));$$

$$\int_0^T (d\langle x \rangle_s(\omega) z(s, \omega),\ z(s, \omega)) =$$

$$= \lim_{n \to \infty} \int_0^T (d\langle x \rangle_s(\omega) z_n(s, \omega),\ z_n(s, \omega)) \tag{4.6}$$

(in probability). If, moreover, we have another sequence $z_n'(t, \omega)$ such that (4.5) holds and the sequence has a limit $z'(t, \omega)$ in the above sense, then we put

$$\int_0^T (d\langle x \rangle_s(\omega) z(s, \omega),\ z'(s, \omega)) =$$

$$= \lim_{n \to \infty} \int_0^T (d\langle x \rangle_s(\omega) z_n(s, \omega),\ z_n'(s, \omega)).$$

Using this extension of the integral, we can define it for a larger class of functions, with properties I-III remaining valid (the passage to the limit in (4.6) is simultaneous for all admissible values of T by Theorem 1 of § 9, and the limit is a continuous local martingale).

The greatest difficulty here is the following: when is it appropriate to assign some limiting value to a sequence $z_n(t, \omega)$ satisfying (4.5)? With the above construction of the function $z(t, \omega)$, the definition of the integral $I(z, x)_t$ may well fail to be single-valued. In the next subsection we shall outline a fairly broad class of functions

$\langle x\rangle_t(\omega)$ for which the integral may indeed be constructed.

10.2. Integrals for Processes with Regular Characteristics

In order to construct a stochastic integral, we must define the integral on the right of (4.3). It follows from Theorem 2 of § 9 that, if $\langle x\rangle_t(\omega)$ is the characteristic of an $\mathbf{X}$-valued martingale, we can define

$$\lambda_t(\omega) = \mathrm{sp}\,\langle x\rangle_t(\omega) = \sum_{k=1}^{\infty} (\langle x\rangle_t(\omega)\, e_k, e_k),$$

where e_k is an orthonormal basis in $\mathbf{X}$; the function $\lambda_t(\omega)$ thus defined is continuous, nondecreasing and $\mathfrak{f}_t$-measurable. It is readily seen that for all $y \in \mathbf{X}$, $t_1 < t_2$,

$$(\langle x\rangle_{t_2}(\omega)\, x, x) - (\langle x\rangle_{t_1}(\omega)\, x, x) \leqslant [\lambda_{t_2}(\omega) - \lambda_{t_1}(\omega)]\,(x, x).$$

Hence there exists a function $\rho(t, \omega, x)$ such that

$$(\langle x\rangle_t(\omega)\, x, x) = \int_0^t \rho(s, \omega, x)\, d\lambda_s(\omega),$$

where it may be assumed that $0 \leqslant \rho(t, \omega, x) \leqslant (x, x)$ and $\rho(t, \omega, x)$ is $\mathfrak{f}_t$-measurable. Setting $\rho(t, \omega, x, y) = \frac{1}{2}[\rho(t, \omega, x+y) - \rho(t, \omega, x) - \rho(t, \omega, y)]$, one sees that there is an operator $R_t(\omega)$ in $\mathbf{L}_w(\Omega, \mathbf{X})$ such that

$$\rho(t, \omega, x, y) = (R_t(\omega)\, x, y),$$

and moreover it may be stipulated that $R_t(\omega)$ is a nonnegative symmetric operator, with $(R_t(\omega)\, x, x) \leqslant (x, x)$. Hence it follows that $R_t(\omega) \in \mathbf{L}(\Omega, \mathbf{X})$ and $\| R_t(\omega) \| \leqslant 1$. Thus,

$$\langle x\rangle_t(\omega) = \int_0 R_s(\omega)\, d\lambda_s(\omega),$$

where $R_s(\omega) \in \mathbf{L}(\Omega, \mathbf{X})$ and $R_s(\omega)$ is symmetric and nonnegative. Let $R_s^{1/2}(\omega)$ be a positive square root of $R_s(\omega)$. Obviously, $\| R_s^{1/2}(\omega) \| \leqslant 1$, and $R_s^{1/2}(\omega)$ is also $\mathfrak{f}_t$-measurable. The characteristic $\langle x\rangle_t(\omega)$ admits the representation

$$\langle x\rangle_t(\omega) = \int_0^t R_s^{1/2}(\omega)\, R_s^{1/2}(\omega)\, d\lambda_s(\omega),$$

which will be used to construct the integral on the right of (4.3). We now consider a larger class of operator-valued functions $\langle x\rangle_t(\omega)$, which we shall need in the sequel. Let us say that an operator-valued function with values in $\mathbf{L}_w(\Omega, \mathbf{X})$ is regular if it may be expressed in the form

$$\langle x\rangle_t(\omega) = \int_0^t (\Phi_s(\omega)\, \Phi_s(\omega)\, d\lambda_s(\omega), \tag{4.7}$$

where $\Phi_s(\omega) \in \mathbf{L}_s(\Omega, \mathbf{X})$ and $\Phi_s^*(\omega) \in \mathbf{L}_s(\Omega, \mathbf{X})$, $\lambda_s(\omega)$ is a continuous nondecreasing function, $|\Phi_s(\omega)\, x|$ is bounded with respect to s, and $(\Phi_s(\omega)\, x, y)$ and $\lambda_s(\omega)$ are $\mathfrak{f}_s$-measurable for $x, y \in \mathbf{X}$. The integral on the right of (4.7)($x, y \in \mathbf{X}$) is defined in the weak sense by the formula

$$(\langle x\rangle_t(\omega)\, x, y) = \int_0^t (\Phi_s(\omega)\, x,\ \Phi_s(\omega)\, y)\, d\lambda_s(\omega). \tag{4.8}$$

It is assumed that for all $x \in \mathbf{X}$ the function $\Phi_s(\omega)\, x$ is measurable with respect to $\mathfrak{A}_+ \times \mathfrak{S}$ for all $t > 0$

$$\int_0^t |\Phi_s(\omega)\, x|^2\, d\lambda_s(\omega) < \infty$$

with probability 1.

Let $\mathcal{F}_{\langle x\rangle}$ denote the set of functions $z(t, \omega)$ in $\mathcal{F}$ such that, for all $T > 0$,

$$\int_0^T |\Phi_s(\omega)\, z(s, \omega)|^2\, d\lambda_s(\omega) < \infty \tag{4.9}$$

with probability 1, where

$$|\Phi_s(\omega)\, z(s, \omega)|^2 = \sum_{k=1}^{\infty} (\Phi_s(\omega)\, z(s, \omega), e_k)^2 =$$

$$= \sum_{k=1}^{\infty} (z(s, \omega),\ \Phi_s^*(\omega)\, e_k)^2$$

(the right-hand side of this relationship is defined thanks to the existence of $\Phi_s^*(\omega)\, x$ for all $x \in \mathbf{X}$). If $z(s, \omega)$ and

$z'(s, \omega)$ are functions in $\mathcal{F}_{\langle x \rangle}$, we put

$$\int_0^T (d\langle x\rangle_s(\omega)\, z(s, \omega),\ z'(s, \omega)) =$$

$$= \int_0^T (\Phi(s, \omega)\, z(s, \omega),\ \Phi(s, \omega)\, z'(s, \omega))\, d\lambda_s(\omega). \qquad (4.10)$$

We also denote $\mathcal{F}^0_{\langle x \rangle} = \mathcal{F}^0 \cap \mathcal{F}_{\langle x \rangle}$.

Theorem. For any function $z(t, \omega) \in \mathcal{F}_{\langle x \rangle}$, there exists a sequence $z_n(t, \omega)$ of functions in $\mathcal{F}^0_{\langle x \rangle}$ such that, for all $T > 0$,

$$\lim_{n\to\infty} \int_0^T |\Phi_s(\omega)\, z(s, \omega) - \Phi_s(\omega)\, z_n(s, \omega)|^2\, d\lambda_s(\omega) = 0 \quad (4.11)$$

(in probability).

Proof. It will suffice to prove the existence of a sequence $z_n(t, \omega)$ such that (4.11) holds for a given T. Indeed, if $z_n^{(k)}(t, \omega)$ is a sequence satisfying (4.11) for $T = k$, then the required sequence is obtained by setting $z_n(t, \omega) = z_n^{(k)}(t, \omega)$ for $k - 1 < t < k$. Let T be fixed. Choose a basis $\{e_k\}$. For any k, $|\Phi_s(\omega)\, e_k|$ is a bounded function of s; hence one can select nonrandom $\alpha_k \downarrow 0$ so that the function

$$\sum_{k=1}^{\infty} \alpha_k |\Phi_s(\omega)\, e_k|$$

is also bounded on $[0, T]$. Let

$$\varphi_s(\omega) = \sup_{\tau < s} \sum \alpha_k |\Phi_\tau(\omega)\, e_k|, \quad \chi_n(t, \omega) = \begin{cases} 1, & \varphi_t(\omega) \leqslant n; \\ 0, & \varphi_t(\omega) > n. \end{cases}$$

It is clear that in order to obtain a sequence satisfying (4.11) it will suffice to construct, for any l, a sequence of functions $z_n^{(l)}(t, \omega) \in \mathcal{F}^0$ such that

$$\lim_{n\to\infty} \int_0^T |\Phi_s(\omega)\, z(s, \omega) - \Phi_s(\omega)\, z_n^{(l)}(s, \omega)|^2\, \chi_l(s, \omega)\, d\lambda_s(\omega) = 0,$$

since, if the sequence $l_n \uparrow \infty$ is chosen in such a way that

$$\lim_{n\to\infty}\int_0^T |\Phi_s(\omega) z(s,\omega) - \Phi_s(\omega) z_n^{l_n}(s,\omega)|^2 \chi_{l_n}(s,\omega)\, d\lambda_s(\omega) = 0,$$

it is also true that

$$\lim_{n\to\infty}\int_0^T |\Phi_s(\omega) z(s,\omega) - \Phi_s(\omega) z_n^{l_n}(s,\omega)|^2 d\lambda_s(\omega) = 0$$

thanks to the inequality

$$\mathbf{P}\left\{\int_0^T |\Phi_s(\omega) z(s,\omega) - \Phi_s(\omega) z_n^{l_n}(s,\omega)|^2 d\lambda_s(\omega) \neq \right.$$

$$\left. \neq \int_0^T |\Phi_s(\omega) z(s,\omega) - \Phi_s(\omega) z_n^{l_n}(s,\omega)|^2 \chi_{l_n}(s,\omega)\, d\lambda_s(\omega)\right\} \leqslant$$

$$\leqslant \mathbf{P}\{\chi_{l_n}(T,\omega) = 0\} \to 0 \quad \text{as } n\to\infty.$$

Thus, we can replace $\Phi_s(\omega)$ in (4.11) by $\chi_l(s,\omega)\Phi_s(\omega)$. Therefore, without loss of generality, we may assume that $|\Phi_s(\omega) e_k|$ is bounded, for any k, by a nonrandom constant. Since it follows from (4.9) that

$$\lim_{n\to\infty}\int_0^T |\Phi_s(\omega) z(s,\omega)|^2 \chi_{\{|\Phi(s,\omega)z(s,\omega)|+|z(s,\omega)|>n\}}(s)\, d\lambda_s(\omega) = 0,$$

the function $z(s,\omega)$ may be so chosen that $|\Phi(s,\omega) z(s,\omega)|$ and $|z(s,\omega)|$ are bounded by nonrandom constants. We may assume that $\lambda_T(\omega)$ is bounded, since (4.11) is equivalent to

$$\lim_{n\to\infty}\int_0^T |\Phi_s(\omega) z(s,\omega) - \Phi_s(\omega) z_n(s,\omega)|^2 d\bar{\lambda}_s(\omega) = 0,$$

where

$$\bar{\lambda}_t(\omega) = \int_0^t e^{-\lambda_s(\omega)} d\lambda_s(\omega), \text{ a } \int_0^T e^{-\lambda_s(\omega)} d\lambda_s(\omega) \leqslant 1.$$

Thus, we may assume that

$$\mathbf{M}\int_0^T |\Phi_s(\omega) z(s, \omega)|^2 d\lambda_s(\omega) < \infty, \tag{4.12}$$

and in order to prove (4.11) it will suffice to show that, under this assumption, there exists a sequence such that

$$\lim_{n\to\infty} \mathbf{M}\int_0^T |\Phi_s(\omega) z(s, \omega) - \Phi_s(\omega) z_n(s, \omega)|^2 d\lambda_s(\omega) = 0. \tag{4.13}$$

Let $z_n(s, \omega)$ be a sequence of step functions such that

$$\lim_{n\to\infty} \mathbf{M}\int_0^T |\Phi_s(\omega) z(s, \omega) - \Phi_s(\omega) z_n(s, \omega)|^2 d\lambda_s(\omega) =$$

$$= \inf_{u\in\mathcal{F}^0} \mathbf{M}\int_0^T |\Phi_s(\omega) z(s, \omega) - \Phi_s(\omega) u(s, \omega)|^2 d\lambda_s(\omega). \tag{4.14}$$

The set of functions in $\mathcal{F}_{\langle x\rangle}$ for which (4.12) is valid may be viewed as a pre-Hilbert space with scalar product

$$(z, z') = \mathbf{M}\int_0^T (\Phi_s(\omega) z(s, \omega),\ \Phi_s(\omega) z'(s, \omega))\, d\lambda_s(\omega)$$

(more precisely, one should first identify functions such that

$$\mathbf{M}\int_0^T |\Phi_s(\omega) z(s, \omega) - \Phi_s(\omega) z'(s, \omega)|^2 d\lambda_s(\omega) = 0\ )$$

Hence our sequence $z_n(s, \omega)$ is fundamental:

$$\lim_{n,m\to\infty} \mathbf{M}\int_0^T |\Phi_s(\omega) z_n(s, \omega) - \Phi_s(\omega) z_m(s, \omega)|^2 d\lambda_s(\omega) = 0 \tag{4.15}$$

and for any function $v(s, \omega) \in \mathcal{F}^0$ such that

$$\mathbf{M}\int_0^T |\Phi_s(\omega) v(s, \omega)|^2 d\lambda_s(\omega) < \infty, \tag{4.16}$$

we have

$$\lim_{n\to\infty} \mathbf{M} \int_0^T (\Phi_s(\omega)[z(s, \omega) - z_n(s, \omega)],\ \Phi(s, \omega) v(s, \omega)\, d\lambda_s(\omega) = 0. \tag{4.17}$$

From (4.14) and (4.15) we deduce that there exists an $\mathbf{X}$-valued process $y(s, \omega)$, such that

$$\lim_{n\to\infty} \mathbf{M} \int_0^T | y(s, \omega) - \Phi_s(\omega)[z(s, \omega) - z_n(s, \omega)] |^2\, d\lambda_s(\omega) = 0, \tag{4.18}$$

and from (4.17) and (4.18) that for all $v \in \mathcal{F}^0$ satisfying (4.11)

$$\mathbf{M} \int_0^T (y(s, \omega),\ \Phi_s(\omega) v(s, \omega))\, d\lambda_s(\omega) = 0. \tag{4.19}$$

Let A be a symmetric positive operator such that

$$A\Phi_s^*(\omega) y(s, \omega) \in \mathbf{X}(\Omega),\ \mathbf{M} \int_0^T | A\Phi^*(\omega) y(s, \omega) |^2\, d\lambda_s(\omega) < \infty.$$

Since

$$A\Phi_s^*(\omega) y(s, \omega) = \sum_{k=1}^{\infty} (A\Phi_s^*(\omega) y(s, \omega),\ e_k) e_k =$$

$$= \sum_{k=1}^{\infty} (y(s, \omega),\ \Phi_s(\omega) A e_k) e_k,$$

it follows that, putting $Ae_k = \rho_k e_k$, we obtain the required operator if

$$| A\Phi^* y(s, \omega) |^2 = \sum_{k=1}^{\infty} (y(s, \omega),\ \Phi_s(\omega) e_k)^2 \rho_k^2 \leqslant$$

$$\leqslant \sum_{k=1}^{\infty} | y(s, \omega) |^2 | \Phi_s(\omega) e_k |^2 \rho_k^2 < \infty,$$

and this may be ensured by the choice of $\rho_k > 0$, since the numbers $| \Phi_s(\omega) e_k |$ are bounded in probability.

It then follows from (4.19) that

$$\mathbf{M}\int_0^T (A\Phi_s^*(\omega)\, y(s,\omega),\ v_1(s,\omega))\, d\lambda_s(\omega) = 0 \qquad (4.20)$$

for any $v_1(s,\omega)$ such that

$$\mathbf{M}\int_0^T |v_1(s,\omega)|^2\, d\lambda_s(\omega) < \infty. \qquad (4.21)$$

But then $A\Phi_s^*(\omega)\, y(s,\omega) = 0$ almost everywhere in the measure π defined on $\mathfrak{A}_+ \times \mathfrak{S}$ by

$$\pi(\Delta \times \Gamma) = \mathbf{M}\chi_\Gamma(\omega)\int_\Delta d\lambda_s(\omega).$$

Therefore,

$$(A\Phi_s^*(\omega)\, y(s,\omega),\ e_k) = (y(s,\omega),\ \Phi_s(\omega)\, e_k)\, \rho_k = 0$$

almost everywhere in π. Thus, we can put $A = I$, since

$$\Phi_s^*(\omega)\, y(s,\omega) = \sum_{k=1}^{\infty} (y(s,\omega),\ \Phi_s(\omega)\, e_k)\, e_k = 0 \in \mathbf{X},$$

so that, for any function $v(s,\omega)$ satisfying (4.21)

$$0 = \mathbf{M}\int_0^T (\Phi_s^*(\omega)\, y(s,\omega),\ v(s,\omega))\, d\lambda_s(\omega) =$$

$$= \mathbf{M}\int_0^T (y(s,\omega),\ \Phi_s(\omega)\, v(s,\omega))\, d\lambda_s(\omega) = 0.$$

In particular, we can put $v(s,\omega) = z(s,\omega)$, and then

$$\mathbf{M}\int_0^T (y(s,\omega),\ \Phi_s(\omega)\, z(s,\omega))\, d\lambda_s(\omega) = 0.$$

This equality, together with (4.19) (with $v \in \mathcal{F}^0$), gives

$$\mathbf{M}\int_0^T (y(s,\omega),\ \Phi_s(\omega)\,[z(s,\omega) - z_n(s,\omega)])\, d\lambda_s(\omega) = 0. \qquad (4.22)$$

It now follows from (4.17) and (4.22) that

$$\mathbf{M}\int_0^T (y(s, \omega), y(s, \omega))\, d\lambda_s(\omega) = 0. \qquad (4.23)$$

But then, by virtue of (4.18),

$$\overline{\lim_{n\to\infty}}\, \mathbf{M}\int_0^T |\Phi_s(\omega)[z(s, \omega) - z_n(s, \omega)]|^2\, d\lambda_s(\omega) \leqslant$$

$$\leqslant \overline{\lim_{n\to\infty}}\, 2\mathbf{M}\int_0^T |y(s, \omega) - \Phi_s(\omega)[z(s, \omega) - z_n(s, \omega)]|^2\, d\lambda_s(\omega) +$$

$$+ 2\mathbf{M}\int_0^T |y(s, \omega)|^2\, d\lambda_s(\omega) = 0.$$

Thus $z_n(s, \omega)$ is the sequence whose existence is claimed in the assertion of our theorem. ∎

Let $x(t, \omega)$ be a continuous local $\mathbf{X}$-valued martingale whose characteristic may be represented in the form (4.7), and let $\Phi_s(\omega)$ and $\lambda_s(\omega)$ satisfy the conditions listed at the beginning of this subsection. Then for all $z(t, \omega)$ we can define a stochastic integral

$$\int_0^T (z(t, \omega), dx(t, \omega)),$$

which is a continuous local martingale as a function of T; its characteristic is

$$\int_0^T (\Phi_s(\omega) z(s, \omega), \Phi_s(\omega) z(s, \omega))\, d\lambda_s(\omega),$$

and the joint characteristic of two such martingales is given by the formula

$$\langle I(z, x), I(z', x)\rangle_t = \int_0^t (\Phi_s(\omega) z(s, \omega), \Phi_s(\omega) z'(s, \omega))\, d\lambda_s(\omega).$$

The existence of the integral follows from the theorem. If $z_n(t, \omega) \in \mathcal{F}^0$ and (4.11) holds, then

$$\lim_{n,m\to\infty}\int_0^T |\Phi_s(\omega)\, z_n(s,\ \omega) - \Phi_s(\omega)\, z_m(s,\ \omega)|^2\, d\lambda_s(\omega) = 0.$$

Therefore, as indicated in subsection 10.1, the limit

$$\lim \int_0^T (z_n(t,\ \omega),\ dx(t,\ \omega)),$$

exists and is the required integral.

10.3. Stochastic Integral with respect to a Wiener Process

Let $w(t,\ \omega)$ be an $\mathbf{X}$-valued Wiener process such that $\mathbf{M}(w(t,\ \omega),\ x) = 0$ and $\mathbf{M}(w,\ (t,\ \omega),\ x)^2 = t\,(Bx,\ x)$, where B is a bounded operator. Then $w(t,\ \omega)$ is an $\mathbf{X}$-valued martingale with characteristic tB. Since

$$tB = \int_0 B^{1/2} B^{1/2} ds,$$

this characteristic is regular. The class $\mathcal{F}_{\langle w\rangle}$ is exactly the set of all functions $z(t,\ \omega)$ such that

$$\int_0^T |B^{1/2} z(t,\ \omega)|^2 dt = \int_0^T (Bz(t,\ \omega),\ z(t,\ \omega))\, dt < \infty.$$

For all $z \in \mathcal{F}_{\langle w\rangle}$ the integral

$$\int_0^T (z(t,\ \omega),\ dw(t,\ \omega))$$

is defined, and if also $z' \in \mathcal{F}_{\langle w\rangle}$, then the joint characteristic of the first integral and $\int_0^T (z'(t,\ \omega),\ dw(t))$ is $\int_0^T (Bz(t,\ \omega),\ z'(t,\ \omega))\, dt.$

11. Stochastic Integral with respect to an Operator-Valued Martingale

11.1. Integrals of X-valued Functions

Let $Y_t(\omega)$ be a continuous operator-valued martingale adapted to a filtration of σ-algebras $\mathfrak{f}_t$, and $\langle Y\rangle_t(\omega)$ its characteristic (see subsection 3, § 9). Let us assume that this martingale is regular in the sense of subsection 2, § 10, i.e.,

$$\langle Y\rangle_t(\omega) = \int_0^t \Phi_s^*(\omega)\,\Phi_s(\omega)\,d\lambda_s(\omega), \tag{4.24}$$

where $\Phi_s(\omega)$ and $\Phi_s^*(\omega) \in \mathbf{L}_s(\Omega, \mathbf{X})$, $\Phi_s(\omega)\,x$ and $\lambda_s(\omega)$ are measurable relative to $\mathfrak{A}_+ \times \mathfrak{S}$, $|\Phi_s(\omega)\,x|$ and $\lambda_s(\omega)$ are continuous in s with probability 1; in addition, $\lambda_s(\omega)$ is a nondecreasing function. We are going to construct a stochastic integral for all functions $z(s, \omega) \in \mathcal{F}_{\langle Y\rangle}$ (see § 10),

$$I(Y, z)_t = \int_0^t (dY_s(\omega),\ z(s, \omega)), \tag{4.25}$$

satisfying the following conditions:

I. $I(Y, z)_t$ is a continuous $\mathbf{X}$-valued local martingale adapted to the filtration $\mathfrak{f}_t$.

II. The joint characteristic of continuous $\mathbf{X}$-valued local martingales $I(Y, z)_t$ and $I(Y, z')_t$ $(z, z' \in \mathcal{F}_{\langle Y\rangle})$ is defined by the formula

$$\langle I(Y, z),\ I(Y, z')\rangle_t = \int_0^t (d\,\langle Y\rangle_s(\omega)\,z(s, \omega),\ z'(s, \omega)) =$$

$$= \int_0^t (\Phi(s, \omega)\,z(s, \omega),\ \Phi(s, \omega)\,z'(s, \omega)\,d\lambda_s(\omega) \tag{4.26}$$

(the integrals on the right of (4.26) where defined in § 10, subsection 2).

III. If $z, z' \in \mathcal{F}_{\langle Y\rangle}$, $\alpha, \alpha' \in \mathbf{R}$, then with probability 1

$$I(Y,\ \alpha z + \alpha' z')_t = \alpha I(Y,\ z)_t + \alpha' I(Y,\ z')_t.$$

Our construction of the stochastic integral (4.25) will be analogous to that of the integral in the preceding section.

Let $z(s, \omega) \in \mathcal{F}^0_{\langle Y \rangle}$, i.e., a step function in $\mathcal{F}_{\langle Y \rangle}$. If $z(t, \omega) = z(t_k, \omega)$ for $t_k < t \leqslant t_{k+1}$, where $t_k \uparrow \infty$, then we put $(t_m < t \leqslant t_{m+1})$

$$\int_0^t dY_s(\omega) z(s, \omega) = \sum_{k=0}^{m-1} (Y_{t_{k+1}}(\omega) - Y_{t_k}(\omega)) z(t_k, \omega) +$$

$$+ (Y_t(\omega) - Y_{t_m}(\omega)) z(t_m, \omega). \tag{4.27}$$

To verify that the expression on the right of (4.27) is meaningful, we prove the following lemma.

Lemma 1. Let $Y_t(\omega)$ be a continuous operator-valued martingale adapted to $\mathfrak{f}_t$, with characteristic $\langle Y \rangle_t(\omega)$ satisfying (4.24); let $z(\omega)$ be an $\mathbf{X}$-valued $\mathfrak{f}_{t_0}$-measurable r.v. such that

$$\mathbf{P}\{([\langle Y \rangle_{t_1}(\omega) - \langle Y \rangle_{t_0}(\omega)] z(\omega), z(\omega)) < \infty\} = 1.$$

Then $[Y_{t_1}(\omega) - Y_{t_0}(\omega)] z(\omega)$ is defined as the limit in probability of $\lim_{n\to\infty} [Y_{t_1}(\omega) - Y_{t_0}(\omega)] z_n(\omega)$, where $z_n(\omega)$ is a sequence of finite-valued $\mathfrak{f}_{t_0}$-measurable r.v.'s such that

$$\lim_{n\to\infty} ((\langle Y \rangle_{t_1}(\omega) - \langle Y \rangle_t(\omega)(z(\omega) - z_n(\omega)), (z(\omega) - z_n(\omega))) = 0 \tag{4.28}$$

(in probability); the left-hand side of (4.28) should be understood as

$$\int_{t_0}^{t_1} |\Phi_s(\omega)[z(\omega) - z_n(\omega)]|^2 d\lambda_s(\omega).$$

Proof. Suppose we have constructed a sequence $z_n(\omega)$ of $\mathfrak{f}_{t_0}$-measurable finite-valued r.v.'s satisfying (4.28). Then for $t > t_0$ the expression

$$[Y_t(\omega) - Y_{t_0}(\omega)] z_n(\omega)$$

is defined, since for any finite-valued function

$$\hat{z}(\omega) = \sum_k \chi_k(\omega) x_k \quad (\chi_i(\omega)\chi_k(\omega) = 0 \text{ as } i \neq k)$$

$$[Y_t(\omega) - Y_{t_0}(\omega)]\,\hat{z}(\omega) = \sum_k \chi_k(\omega)\,[Y_t(\omega) - Y_{t_0}(\omega)]\,x_k.$$

It is readily seen that $[Y_t(\omega) - Y_{t_0}(\omega)]\,z(\omega)$ is a continuous local $\mathbf{X}$-valued martingale (thanks to the $\mathfrak{f}_{t_0}$-measurability of $\chi_k(\omega)$), and its scalar quadratic characteristic is

$$\sum_k \chi_k(\omega)\,([\langle Y\rangle_t(\omega) - \langle Y\rangle_{t_0}(\omega)]\,x_k,\ x_k) = ((\langle Y\rangle_t(\omega) -$$

$$-\langle Y\rangle_{t_0}(\omega))\,\hat{z}(\omega),\ \hat{z}(\omega)).$$

Therefore $y_n(t,\omega) = [Y_t(\omega) - Y_{t_0}(\omega)]\,z_n(\omega)$ is a sequence of martingales such that

$$\lim_{n,m\to\infty} \sum_{k=1}^{\infty} (\langle y_n - y_m\rangle_{t_1}(\omega)\,e_k,\ e_k) = \lim_{n,m\to\infty} ([\langle Y\rangle_{t_1}(\omega) -$$

$$-\langle Y\rangle_{t_0}(\omega)]\,(z_n(\omega) - z_m(\omega)),\ z_n(\omega) - z_m(\omega)) = 0$$

(in probability) thanks to (4.28). Hence by Theorem 3 of § 9, the limit $\lim\limits_{n\to\infty} [Y_t(\omega) - Y_{t_0}(\omega)]\,z_n(\omega)$ exists. Thus, to prove the lemma it will suffice to prove that there exists a sequence $z_n(\omega)$, satisfying (4.28).

It follows from the condition

$$([\langle Y\rangle_{t_1}(\omega) - \langle Y\rangle_{t_0}(\omega)]\,z(\omega),\ z(\omega)) = \int_{t_0}^{t_1} |\,\Phi_s(\omega)\,z(\omega)\,|^2\,d\lambda_s(\omega) < \infty$$

that

$$\int_{t_0}^{t} |\,\Phi_s(\omega)\,z(\omega)\,|^2\,d\lambda_s(\omega)$$

is a continuous (and nondecreasing) function of t on $[t_0,\ t_1]$. Put $\lambda_t^{(m)}(\omega) = \lambda_t(\omega)$ if $\int_{t_0}^{t} |\,\Phi_s(\omega)\,z(\omega)\,|^2\,d\lambda_s(\omega) \leqslant m;$

$$\lambda_t^{(m)}(\omega) = \lambda_\tau(\omega) \quad \text{if } \tau \leqslant t \text{ and } \int_{t_0}^{\tau} |\Phi_s(\omega) z(\omega)|^2 \, d\lambda_s(\omega) = m.$$

It will suffice to prove that for any m there exists an $\mathfrak{f}_{t_0}$-measurable sequence of finite-valued functions $z_n^m(\omega)$ such that

$$\lim_{n\to\infty} \mathbf{M} \int_{t_1}^{t} |\Phi_s(\omega) z(\omega) - \Phi_s(\omega) z_n^{(m)}(\omega)|^2 \, d\lambda_s^{(m)}(\omega) = 0. \quad (4.29)$$

Proceeding along lines analogous to those followed in the proof of the theorem of § 10, one establishes the existence of a sequence of finite-valued $\mathfrak{f}_{t_0}$-measurable functions $\hat{z}_n(\omega)$ such that

$$\lim_{n,l\to\infty} \mathbf{M} \int_{t_0}^{t} |\Phi_s(\omega) \hat{z}_n(\omega) - \Phi_s(\omega) \hat{z}_t(\omega)|^2 \, d\lambda_s^{(m)}(\omega) = 0, \quad (4.30)$$

and for any finite-valued $\mathfrak{f}_{t_0}$-measurable function $\hat{z}(\omega)$

$$\lim_{l\to\infty} \mathbf{M} \int_{t_0}^{t} (\Phi_s(\omega) z_n(\omega) - \Phi_s(\omega) \hat{z}_l(\omega), \; \Phi_s(\omega) \hat{z}(\omega)) \times$$

$$\times \, d\lambda_s^{(m)}(\omega) = 0. \quad (4.31)$$

Let $y(s, \omega)$ be a function such that

$$\lim_{n,l\to\infty} \mathbf{M} \int_{t_0}^{t} |y(s, \omega) - \Phi_s(\omega) \hat{z}_l(\omega)|^2 \, d\lambda_s^{(m)}(\omega) = 0$$

(the existence of this function follows from (4.30)). Then it follows from (4.31) that, $\forall z \in \mathbf{X}$,

$$\mathbf{M}\left[\int_{t_0}^{t} (\Phi_s(\omega) z(\omega) - y(s, \omega), \; \Phi_s(\omega) z) \, d\lambda_s^{(m)}(\omega) / \mathfrak{f}_{t_0}\right] = 0, \quad (4.32)$$

provided that

$$\mathbf{M}\int_{t_0}^{t} |\Phi_s(\omega) z|^2 \, d\lambda_s^{(m)}(\omega) < \infty. \tag{4.33}$$

Obviously, we can insert $\mathfrak{F}_{t_0}$-measurable random variables z in (4.32) provided they satisfy (4.33). Hence (4.32) is valid for $z = z(\omega)$ and $z = \hat{z}_l(\omega)$, so that

$$\mathbf{M}\int_{t_0}^{t} (\Phi_s(\omega) z(\omega) - y(s, \omega), \; \Phi_s(\omega) z(\omega) - \Phi_s(\omega) \hat{z}_l(\omega)) \times$$

$$\times \, d\lambda_s^{(m)}(\omega) = 0.$$

Letting $l \to \infty$, we see that

$$\mathbf{M}\int_{t_0}^{t} |\Phi_s(\omega) z(\omega) - y(s, \omega)|^2 \, d\lambda_s^{(m)}(\omega) = 0,$$

i.e.,

$$\lim_{l\to\infty} \mathbf{M}\int_{t_0}^{t} |\Phi_s(\omega) z(\omega) - \Phi_s(\omega) \hat{z}_l(\omega)| \, d\lambda_s^{(m)}(\omega) = 0. \; \blacksquare$$

Remark. In the course of the proof it was shown that, under the assumptions of the lemma($t > t_0$),

$$[Y_t(\omega) - Y_{t_0}(\omega)] \, z(\omega)$$

is a continuous local $\mathbf{X}$-valued martingale, and the joint characteristic of the martingales $[Y_t(\omega) - Y_{t_0}(\omega)] z(\omega)$ and $[Y_t(\omega) - Y_{t_0}(\omega)] z_1(\omega)$ is defined by

$$([\langle Y\rangle_t(\omega) - \langle Y\rangle_{t_0}(\omega)] \, z(\omega), \, z_1(\omega)) =$$

$$= \int_{t_0}^{t} (\Phi_s(\omega) z(\omega), \; \Phi_s(\omega) z_1(\omega)) \, d\lambda_s(\omega) \tag{4.34}$$

Hence, if $\sigma(Y, z, t)$ is the operator characteristic of the local martingale $[Y_t(\omega) - Y_{t_0}(\omega)] z(\omega)$, then

$$\sum_{k=1}^{\infty} (\sigma(Y, z, t) e_k, e_k) = \int_{t_0}^{t} |\Phi_s(\omega) z(\omega)|^2 d\lambda_s(\omega).$$

Thus, the right-hand side of (4.27) is an $\mathbf{X}^{\boldsymbol{\cdot}}$-valued continuous local martingale. For functions $z(s, \omega)$ and $z'(s, \omega)$ in $\mathcal{F}^0_{\langle Y\rangle}$, the relationship (4.34) implies the formula (4.26). We have thus established that the stochastic integral defined by (4.27) satisfies conditions I-III.

If $z_n(t, \omega)$ is a sequence of functions in $\mathcal{F}^0_{\langle Y\rangle}$, then for all $T > 0$

$$\lim_{n\to\infty} \int_0^T |\Phi_s(\omega)[z_n(s, \omega) - z(s, \omega)]|^2 d\lambda_s(\omega) = 0, \qquad (4.35)$$

where $z(t, \omega) \in \mathcal{F}_{\langle Y\rangle}$, then

$$\int_0^t dY_s(\omega) z_n(s, \omega)$$

and therefore, by Theorem 3 of § 9, the sequence

$$\lim_{n,m\to\infty} \int_0^T |\Phi_s(\omega)[z_n(s, \omega) - z_m(s, \omega)]|^2 d\lambda_s(\omega) = 0,$$

of $\mathbf{X}$-valued continuous local martingales converges to some $\mathbf{X}$-valued continuous local martingale, which we denote by

$$\int_0^t dY_s(\omega) z(s, \omega) = \lim_{n\to\infty} \int_0^t dY_s(\omega) z_n(s, \omega). \qquad (4.36)$$

It follows from the theorem of § 10 that for all $z(t, \omega) \in \mathcal{F}_{\langle Y\rangle}$ one can exhibit a sequence $z_n(t, \omega) \in \mathcal{F}^0_{\langle Y\rangle}$ satisfying (4.35). Hence formula (4.36) enables one to define the integral (4.25) for all $z(t, \omega) \in \mathcal{F}_{\langle Y\rangle}$. Since conditions I-III hold for functions $z(t, \omega) \in \mathcal{F}^0_{\langle Y\rangle}$, they also hold for functions $z(t, \omega) \in \mathcal{F}^0_{\langle Y\rangle}$ (condition I follows from the definition; condition II is obtained from the Remark to Theorem 3 of § 9, since it need only be verified when $z' = z$, condition III is verified by taking limits in the relevant equality, which is valid for step functions).

11.2. Integrals of Operator-valued Functions

In this subsection, under the same assumptions concerning the process $Y_t(\omega)$, we shall define integrals in the form

$$\int_0^t Z_s(\omega)\, dY_s(\omega) \text{ and} \int_0^t dY_s(\omega)\, Z_s(\omega),$$

where $Z_s(\omega)$ is an operator-valued function. We begin with the second integral. Since $Z_s(\omega)\, x$ is an $\mathbf{X}$ -valued function, we can define the integral

$$\int_0^t dY_s(\omega)\, Z_s(\omega)\, x,$$

provided $Z_s(\omega)\, x \in \mathcal{F}_{\langle Y \rangle}$, i.e.,

$$\int_0^T |\Phi_s(\omega)\, Z_s(\omega)\, x|^2\, d\lambda_s(\omega) < \infty. \qquad (4.37)$$

Let $Z_s(\omega)$ be a measurable function with values in $\mathbf{L}_\omega(\Omega, \mathbf{X})$ (henceforth, whenever speaking of measurable random operator-valued function $Z_t(\omega)$, we shall mean a function such that for all $x, y \in \mathbf{X}$ the real-valued function $(Z_t(\omega)\, x, y)$ is measurable with respect to $\mathfrak{A}_+ \times \mathfrak{S}$); we shall assume that this function is $\mathfrak{f}_s$ -measurable for any s, that $\Phi_s(\omega)\, Z_s(\omega) \in \mathbf{L}_s(\Omega, \mathbf{X})$ and that (4.37) holds for all $x \in \mathbf{X}$. The set of functions Z satisfying these conditions will be denoted by $T_{\langle Y \rangle}$. For $Z \in T_{\langle Y \rangle}$, we define an operator-valued continuous martingale $I_t(Y, Z)$ by putting, for all $x \in \mathbf{X}$,

$$I_t(Y, Z)\, x = I_t(Y, z_x),$$

where $z_x = Z_t(\omega)\, x$. The operator characteristic of this martingale is, by definition,

$$\langle I(Y, Z) \rangle_t = \int_0^t Z_s^*(\omega)\, \Phi_s^*(\omega)\, \Phi_s(\omega)\, Z_s(\omega)\, d\lambda_s(\omega) \qquad (4.38)$$

(the integral on the right is to be understood in the weak sense: for $x, y \in \mathbf{X}$,

$$\left(\int_0^t Z_s^*(\omega)\,\Phi_s^*(\omega)\,Z_s(\omega)\,ds,\ x,\ y\right) = \int_0^t (Z_s^*(\omega)\,\Phi_s^*(\omega)\,\Phi_s(\omega)\,\times$$

$$\times\, Z_s(\omega)\,x,\ y)\,d\lambda_s(\omega) = \int_0^t (\Phi_s(\omega)\,Z_s(\omega)\,x,\ \Phi_s(\omega)\,Z_s(\omega)\,y)\,\times$$

$$\times\, d\lambda_s(\omega)$$

- its existence is ensured by condition (4.37)).

Denote

$$I_t^*(Z,\ Y) = \int_0^t Z_s(\omega)\,dY_s(\omega). \qquad (4.39)$$

It is easily seen that for step functions $Z_s(\omega)$, defining the right-hand side of (4.39) in the natural manner as a sum

$$\Sigma\, Z_{t_k}(\omega)\,[Y_{t_{k+1}\wedge T}(\omega) - Y_{t_k\wedge T}(\omega)]$$

we have the identity

$$I_t^*(Z,\ Y) = [I_t(Y^*,\ Z^*)]^* \qquad (4.40)$$

(where the asterisk on the right denotes adjoints), provided that the right-hand side is meaningful. We can therefore define the integral (4.39) via (4.40), provided the following conditions hold:

(a) Y_t^* is a continuous local operator-valued martingale with regular characteristic, admitting a representation

$$\langle Y^*\rangle_t(\omega) = \int_0^t \Psi_s^*(\omega)\,\Psi_s(\omega)\,d\lambda_s(\omega);$$

(b) the measurable function $Z_s(\omega)$ has values in $\mathbf{L}_w(\Omega, \mathbf{X})$, is such that $\Psi_s(\omega)\,Z_s^*(\omega) \in \mathbf{L}_s(\Omega, \mathbf{X})$, and

$$\int_0^t Z_s(\omega)\,\Psi_s^*(\omega)\,\Psi_s(\omega)\,Z_s^*(\omega)\,d\lambda_s(\omega) \in \mathbf{L}_w^+(\Omega,\ \mathbf{X}). \qquad (4.41)$$

Then formula (4.40) defines a continuous local martingale with values in $\mathbf{L}_{\omega.}(\Omega, \mathbf{X})$; and in that situation $[I_t^*(Z, Y)]^*$ is a strong operator-valued martingale whose characteristic is defined by the left-hand side of (4.41).

The above definition is rather inconvenient in that, first, additional conditions are imposed on Y_t; second, the resulting integral is no longer necessarily a continuous martingale - - this is true only of the process obtained by going over to adjoints; in particular, only the characteristic of the last--mentioned process can be evaluated.

We shall define the integral (4.39) for a different class of functions $Z_t(\omega)$, as a limit of integrals of simple functions; this will be done without imposing any additional assumptions on the martingale $Y_t(\omega)$. The class of integrable functions turns out, nonetheless, to be rather small.

Let T^0 denote the set of all functions $Z_t(\omega)$, satisfying the following conditions:

(1) $Z_t(\omega)$ is defined for $t \geqslant 0$, assumes values in $\mathbf{L}(\Omega, \mathbf{X})$ and is $\mathfrak{f}_t$-measurable for all $t \geqslant 0$;

(2) $Z_t(\omega)$ is a step function.

Let $Z_t(\omega)$ be constant on intervals $[t_k, t_{k+1})$, where $0 = t_0 < t_1 < \ldots,\ t_n \uparrow \infty$. Put

$$I^*(Z, Y)_t = \Sigma Z_{t_k}(\omega)[Y_{t_{k+1}\wedge t}(\omega) - Y_{t_k\wedge t}(\omega)] = \tag{4.42}$$

$$= \sum_{k<m} Z_{t_k}(\omega)[Y_{t_{k+1}}(\omega) - Y_{t_k}(\omega)] + Z_{t_m}[Y_t(\omega) - Y_{t_m}(\omega)]$$

for $t \in [t_m, t_{m+1}]$. It is obvious that $Z_{t_m}(\omega)[Y_t(\omega) - Y_{t_m}(\omega)]$ is a continuous local operator-valued martingale. Let us evaluate its characteristic. As follows from the Remark to Theorem 2 of § 9, the characteristic of this martingale at time $t_m + t$ is exactly the limit

$$\lim_{n\to\infty}\sum_{k=0}^{n-1} | Z_{t_m}(\omega)(Y_{\frac{k+1}{n}t+t_m}(\omega) - Y_{\frac{k}{n}t+t_m}(\omega))\, x|^2.$$

This is bounded above by the expression

$$\| Z_{t_m}(\omega)\|^2 \lim_{n\to\infty}\sum_{k=0}^{n-1} |(Y_{\frac{k+1}{n}t+t_m}(\omega) - Y_{\frac{k}{n}t+t_m}(\omega))\, x|^2 =$$

$$= \| Z_{t_m}(\omega)\|^2 \int_{t_m}^{t_m+t} |\Phi_s(\omega)\, x|^2\, d\lambda_s(\omega).$$

Thus, defining $I^*(Z, Y)_t$ by formula (4.42), we obtain

$$\langle I^*(Z, Y)\rangle_t \leqslant \int_0^t \| Z_s(\omega)\|^2 \Phi_s^*(\omega)\Phi_s(\omega)\, d\lambda_s(\omega). \tag{4.43}$$

This formula enables us to extend the definition of the integral to the class $T^{(u)}_{\langle Y\rangle}$ of functions $Z_t(\omega)$ satisfying condition (1), and also the following condition:

(3) for any function $Z_t(\omega) \in T^{(u)}_{\langle Y\rangle}$, one can find a sequence $Z_t^{(n)}(\omega)$ such that, for all $t > 0$ and $x \in \mathbf{X}$,

$$\lim_{n\to\infty} \int_0^t \| Z_s^{(n)}(\omega) - Z_s(\omega)\|^2 \, |\Phi_s(\omega)\, x|^2 \, d\lambda_s(\omega) = 0 \tag{4.44}$$

in the sense of convergence in probability. If $Z_t(\omega) \in T^{(u)}_{\langle Y\rangle}$ and $Z_t^{(n)}(\omega)$ satisfies (4.44), then

$$\lim_{m,n\to\infty} \int_0 \| Z_s^{(m)}(\omega) - Z_s^{(n)}(\omega)\|^2 \, |\Phi_s(\omega)\, x|^2 \, d\lambda_s(\omega) = 0$$

in probability for all $t > 0$ and $x \in \mathbf{X}$; hence, for all $t > 0$ and $x \in \mathbf{X}$,

$$\lim_{n,m\to\infty} (\langle I^*(Z^{(m)} - Z^{(n)}, Y)\rangle_t(\omega)\, x, x) = 0,$$

so that, by Theorem 5 of § 9, $I^*(Z^{(n)}, Y)_t$ converges to a continuous local operator-valued martingale, which we denote by $I^*(Z, Y)_t$. Let $\hat{T}^{(u)}_{\langle Y\rangle}$ denote the smallest class of functions satisfying condition (1) which is closed under convergence in the sense: $Z_t^{(n)}(\omega) \to Z_t(\omega)$, if, for all $t > 0$, $x \in \mathbf{X}$,

$$\int_0^t \| Z_t^{(n)}(\omega) - Z_t(\omega)\|^2 \, |\Phi_s(\omega)\, x|^2 \, d\lambda_s(\omega) \to 0,$$

in probability. It is clear that the definition of $I^*(Z, Y)_t$ extends by continuity to $\hat{T}^{(u)}_{\langle Y\rangle}$; when this is done, $I^*(Z, Y)_t$ is a continuous local operator-valued martingale, and its characteristic satisfies inequality (4.43). In addition, $I^*(Z, Y)_t$ is a linear homogeneous function of $Z \in \hat{T}^{(u)}_{\langle Y\rangle}$. We claim that $\hat{T}^{(u)}_{\langle Y\rangle}$ contains all functions $Z_t(\omega)$ satisfying condition (1), such that

$$\int_0^t \| Z_s(\omega)\|^2 \, |\Phi_s(\omega)\, x|^2 \, d\lambda_s(\omega) < \infty$$

for $t \geqslant 0$ and $x \in \mathbf{X}$, and for all t with the possible exception of a countable set

$$\|Z_s(\omega) - Z_t(\omega)\| \to 0$$

in probability as $s \to t$. Indeed, let $V_t(\omega)$ be defined by $V_t(\omega) = Z_t(\omega)$ if $\|Z_t(\omega)\| \leqslant \alpha$, $V_t(\omega) = 0$ if $\|Z_t(\omega)\| > \alpha$, where α is chosen so that

$$\mathbf{P}\left\{\int_0^\infty \chi_{\{\|Z_t(\omega)\|=\alpha\}} e^{-\lambda_t(\omega)} d\lambda_t(\omega) = 0\right\} = 1. \qquad (4.45)$$

Then $V_t(\omega)$ is continuous in the operator norm in probability for almost all t, i.e.,

$$\|V_s(\omega) - V_t(\omega)\| \to 0$$

in probability as $s \to t$, for almost all t. Put

$V_t^{(n)}(\omega) = V_{\frac{k}{n}}(\omega)$ for $\frac{k}{n} \leqslant t < \frac{k+1}{n}$. Then

$$\lim_{n\to\infty} \int_0^t \|V_s^{(n)}(\omega) - V_s(\omega)\|^2 |\Phi_s(\omega) x|^2 d\lambda_s(\omega) = 0$$

in probability, since $\|V_t^{(n)}(\omega) - V_t(\omega)\|^2 \leqslant 4\alpha^2$ and $\|V_t^{(n)}(\omega) - V_t(\omega)\|^2 \to 0$ for almost all t in the measure $d\lambda_t(\omega)$. Thus, if (4.45) holds, then $V_t(\omega) \in T^{(u)}_{\langle Y\rangle}$. Note that for any sequence α_k

$$\sum_k \mathbf{M} \int_0^\infty \chi_{\{\|Z_t(\omega)\|=\alpha_k\}} e^{-\lambda_t(\omega)} d\lambda_t(\omega) \leqslant 1.$$

and so there is an at most countable set of α for which (4.45) fails to hold. It follows from the relationship

$$\int_0^t \|Z_s(\omega) - V_s(\omega)\|^2 |\Phi_s(\omega) x|^2 d\lambda_s(\omega) =$$
$$= \int_0^t \|Z_s(\omega)\|^2 \chi_{\{\|Z_s(\omega)\|>\alpha\}} |\Phi_s(\omega) x|^2 d\lambda_s(\omega)$$

that as $\alpha \to \infty$ the left-hand side tends to zero, so that

$Z_t(\omega) \in \hat{T}^{(u)}_{\langle Y \rangle}$.

Note that the above definitions of the integral $I^*(Z, Y)$ are not contradictory: if Y and Z are such that both integrals are defined, the integrals are identical. This follows from the fact that they are weak limits of the same integral sums.

11.3. Stochastic Integral with respect to an Operator-valued Wiener Process

Let $W_t(\omega)$ be a strong operator-valued Wiener process (see subsection 4, § 9). Let $\beta(x, y, u, v)$ denote a quadrilinear form on $\mathbf{X}$, such that

$$\mathbf{M}(W_t(\omega)x, y)(W_t(\omega)u, v) = t\beta(x, y, u, v).$$

It follows from subsection 4, § 2, that if $W_t(\omega) \in \mathbf{L}_s(\Omega, \mathbf{X})$ then the series

$$\Sigma\beta(x, e_k, x, e_k) = (Bx, x),$$

where B is a nonnegative bounded symmetric operator, is convergent. When that is true (see subsection 4, § 9),

$$\langle W\rangle_t(\omega) = tB.$$

Using the results of the preceding subsection, one shows that $I(W, Z)_t$ is defined for all $Z_t(\omega)$, which are $\mathfrak{f}_t$ -measurable for any t and assume values in $\mathbf{L}_s(\Omega, \mathbf{X})$, such that, for all $t > 0, x \in \mathbf{X}$,

$$\int_0^t |B^{1/2} Z_s(\omega)x|^2 ds = \int_0^t (BZ_s(\omega)x, Z_s(\omega)x)\, ds =$$

$$= \int_0^t \sum_{k=1}^{\infty} \beta(Z_s(\omega)x, e_k Z_s(\omega)x, e_k)\, ds < \infty.$$

In this case the characteristic of the integral $I^*(Z, W)_t$ can be evaluated exactly, thus permitting extension of the integral to a larger class of functions Z.

Let $Z(\omega)$ be some $\mathfrak{f}_{t_k}$ -measurable operator in $\mathbf{L}(\Omega, \mathbf{X})$. Then, when $t > t_k$, the operator $W_t(\omega) - W_{t_k}(\omega)$ is independent of $Z(\omega)$, and, by the lemma of § 8,

$$Z(\omega)\,[W_t(\omega) - W_{t_k}(\omega)] \in \mathbf{L}_s(\Omega, \mathbf{X}).$$

It is easy to see that this product is a weak local operator--valued martingale. Let us evaluate its characteristic - for the moment, purely formally. Define an operator R_x, by putting $(R_x u, v) = \beta(x, u, x, v)$; R_x is a nonnegative symmetric operator with finite trace: it is a quadratic function of x, and $\mathrm{sp}\, R_x = (Bx, x)$. The joint characteristic of the martingales

$$(Z(\omega)\,[W_t(\omega) - W_{t_k}(\omega)]\,x,\ y); \quad (Z(\omega)\,[W_t(\omega) - W_{t_k}(\omega)]\,x,\ u)$$

is precisely

$$\mathbf{M}\,\{(Z(\omega)\,[W_t(\omega) - W_{t_k}(\omega)]\,x,\ y)\,(Z(\omega)\,[W_t(\omega) - W_{t_k}(\omega)] \times$$
$$\times\, x,\ u)/\mathfrak{f}_{t_k}\} = (t - t_k)\,(R_x Z^*(\omega)\,y,\ Z^*(\omega)\,x) =$$
$$= (t - t_k)\,(Z(\omega)\,R_x Z^*(\omega)\,y,\ u).$$

Note that $Z(\omega)R_x Z^*(\omega)$ is defined as a weak operator, and it is nonnegative and symmetric. Indeed, let $\{f_n(x)\}$ be a basis of eigenvectors of R_x, $\lambda_n(x)$ the corresponding eigen-values. Then

$$(Z(\omega)\,R_x Z^*(\omega)\,y,\ u) = \sum_n (R_x Z^*(\omega)\,y,\ f_n)\,(f_n,\ Z^*(\omega)\,u) =$$
$$= \sum_n \lambda_n(x)\,(Z^*(\omega)\,y,\ f_n)\,(f_n,\ Z^*(\omega)\,u)$$

and so $Z(\omega)\,R_x Z^*(\omega) \in \mathbf{L}_w(\Omega,\ \mathbf{X})$.

Up to an $\mathfrak{f}_{t_k}$-measurable factor, the process $Z(\omega)\,[W_t(\omega) - W_{t_k}(\omega)]$ is a Gaussian process with independent increments; its characteristic is expressed in terms of a conditional expectation:

$$\mathbf{M}\,|\,Z(\omega)\,[W_t(\omega) - W_{t_k}(\omega)]\,x\,|^2 = (t - t_k) \sum_{j=1}^{\infty} (Z(\omega)\,R_x \times$$
$$\times\, Z^*(\omega)\,e_j,\ e_j) = (t - t_k)\,\mathrm{sp}\,Z(\omega)\,R_x Z^*(\omega) = (t - t_k) \times$$
$$\times \sum_{j=1}^{\infty} \sum_{n=1}^{\infty} (Z(\omega)\,f_n(x),\ e_j)^2 = (t - t_k) \sum_{n=1}^{\infty} \lambda_n(x)\,|\,Z(\omega)\,f_n(x)\,|^2.$$

The right-hand side of the last equality is not always de-fined. We shall show that if the series $\sum_{n=1}^{\infty} \lambda_n\,|\,Z(\omega)\,f_n(x)\,|^2$ is convergent, then $Z(\omega)[W_t(\omega) - W_{t_k}(\omega)]$ is a continuous local martingale in $\mathbf{X}$. Indeed, in that case

$$Z(\omega)[W_t(\omega) - W_{t_k}(\omega)]x = \sum_{n=1}^{\infty} \beta_t(n, x, \omega) f_n(x), \qquad (4.46)$$

where

$$\beta_t(n, x, \omega) = (Z(\omega)[W_t(\omega) - W_{t_k}(\omega)]x, f_n(x)) =$$

$$= \sum_{m=1}^{\infty} ([W_t(\omega) - W_{t_k}(\omega)]x, f_m(x))(Z^*(\omega) f_m(x), f_n(x)).$$

Note that $w_m(t) = ([W_t(\omega) - W_{t_k}(\omega)]x, f_m(x))$ is a one-dimensional Wiener process, and

$$\mathbf{M} w_m(t) w_l(t) = (t - t_k)(R_x f_m(x), f_l(x)) = \lambda_m(x)(t - t_k)\delta_{lm}$$

Thus,

$$\beta_t(n, x, \omega) = \sum_{m=1}^{\infty} w_m(t)(Z(\omega) f_m(x), f_n(x)).$$

Substituting this expression into (4.46), we find that

$$Z(\omega)[W_t(\omega) - W_{t_k}(\omega)]x = \sum_{m=1}^{\infty} w_m(t) \sum_{n=1}^{\infty} (Z(\omega) f_m(x), f_n(x)) f_n(x) =$$

$$= \sum_{m=1}^{\infty} w_m(t) Z(\omega) f_m(x).$$

Since $Z(\omega) f_m(x)$ is an $\mathfrak{f}_{t_k}$-measurable $\mathbf{X}$-valued r.v., it follows that for all n

$$\sum_{m=1}^{n} w_m(t) Z(\omega) f_m(x)$$

is a continuous local $\mathbf{X}$-valued martingale with characteristic

$$(t - t_k) \sum_{m=1}^{n} \lambda_m(x) | Z(\omega) f_m(x) |^2.$$

As $n \to \infty$ this expression converges to

$$(t - t_k) \times \sum_{m=1}^{\infty} \lambda_m(x) | Z(\omega) f_m(x) |^2 ;$$

together with Theorem 3 of § 9, this implies that

$\sum_{m=1}^{\infty} w_m(t) Z(\omega) f_m(x)$ converges to a continuous local $\mathbf{X}$-valued martingale with characteristic

$$(t - t_k) \sum_{n=1}^{\infty} \lambda_n(x) \,|\, Z(\omega\; f_n(x) |^2 = (t - t_k) \operatorname{sp} Z(\omega) R_x Z^*(\omega) =$$

$$= (t - t_k) \sum_{n=1}^{\infty} \beta(x, Z^*(\omega) e_k, x, Z^*(\omega) e_k).$$

Let $Z(t, \omega)$ be an $\mathfrak{f}_t$-measurable step function for all t, with values in $\mathbf{L}_s(\Omega, \mathbf{X})$, such that almost surely, for all t and x, the following expression is bounded:

$$\int_0^t \sum_{n=1}^{\infty} \beta(x, Z_s^*(\omega) e_k, x, Z_s^*(\omega) e_k)\, ds. \tag{4.47}$$

The set of all such functions will be denoted by T_β^0. For any $Z \in T_\beta^0$ we can define a stochastic integral $I^*(Z, W)_t$ via (4.42); with this definition, $I^*(Z, W)_t$ will be a continuous local operator-valued martingale, whose characteristic is defined by (4.47).

Let T_β denote the minimal class of functions $Z_t(\omega)$ satisfying the following conditions:

(a) $Z_t(\omega)$ is defined for $t \geqslant 0$, takes values in $\mathbf{L}_s(\Omega, \mathbf{X})$, is jointly measurable in t and ω, and is $\mathfrak{f}_t$-measurable for all t;

(b) T_β contains T_β^0;

(c) if $Z_t^{(n)}(\omega) \in T_\beta$ and $Z_t(\omega)$ are such that for all x, $t > 0$

$$\lim_{n\to\infty} \int_0^t \operatorname{sp} [Z_s(\omega) - Z_s^{(n)}(\omega)] R_x [Z_s(\omega) - Z_s^{(n)}(\omega)]^* ds = 0,$$

then $Z_t(\omega) \in T_\beta$. The stochastic integral $I^*(Z, W_t)$ may be extended to the class T_β by continuity.

We now describe a sufficient condition for a function $Z_s(\omega)$ to be an element of T_β. An operator-valued function $Z_s(\omega)$ with values in $\mathbf{L}_s(\Omega, \mathbf{X})$ is said to be strongly continuous in mean square if, for all $x \in \mathbf{X}$, $\mathbf{M} \,|\, Z_s(\omega) x |^2 < \infty$ and for $t \geqslant 0$

$$\lim_{s\to t} \mathbf{M} \,|\, Z_s(\omega)\, x - Z_t(\omega)\, x \,|^2 = 0. \tag{4.48}$$

Let B_{st} denote the bounded operator defined by

$$\mathbf{M}\,(Z_s(\omega)\, x,\ Z_t(\omega)\, y) = (B_{st}x,\ y)$$

(that this operator is bounded follows from the boundedness of B_{tt}, which in turn follows from the lemma of § 2 and from the inequality $(B_{st}x,\ y)^2 \leqslant (B_{ss}x,\ x)\,(B_{tt}y,\ y))$. Note that $B_{st} = \mathbf{M}Z_t^*(\omega)\, Z_s(\omega)$. The function $(B_{st}\ x,\ y)$ is jointly continuous in s and t, since

$$|\,(B_{st}x,\ y) - (B_{\sigma\tau}x,\ u)\,| = |\, \mathbf{M}\,(Z_\varepsilon(\omega)\, x - Z_\sigma(\omega)\, x,\ Z_t(\omega)\, y) +$$

$$+ \mathbf{M}\,(Z_\sigma(\omega)\, x,\ Z_t(\omega)\, y - Z_\tau(\omega)\, y)\,| \leqslant (\mathbf{M}\,|\, Z_s(\omega)\, x -$$

$$- Z_\sigma(\omega)\, x\,|^2)^{1/2}\, (\mathbf{M}\,|\, Z_t(\omega)\, y\,|^2)^{1/2} + (\mathbf{M}\,|\, Z_t(\omega)\, y - Z_\tau(\omega)\, y\,|^2)^{1/2},$$

and from (4.48) it follows that $\mathbf{M}\,|\, Z_t(\omega)\, x\,|^2$ is bounded on any finite closed interval.

Since B_{st} is continuous, $\|B_{st}\|$ is bounded on any bounded domain in the s, t-plane.

Lemma 2. If the operator-valued function $Z_\cdot(\omega)$ satisfies condition (a) and is strongly continuous in mean square, then $Z_s(\omega) \in T_\beta$.

Proof. Put $Z_s^{(n)}(\omega) = Z_{\frac{[ns]}{n}}(\omega)$, where $[ns]$ is the integral part of ns. Then

$$\mathbf{M} \int_0^t \operatorname{sp}\, [Z_s(\omega) - Z_s^{(n)}(\omega)]\, R_x\, [Z_s(\omega) - Z_s^{(n)}(\omega)]^*\, ds =$$

$$= \int_0^t \operatorname{sp} \mathbf{M}\, [Z_s(\omega) - Z_{\frac{[ns]}{n}}(\omega)]^*\, [Z_s(\omega) - Z_{\frac{[ns]}{n}}(\omega)]\, R_x ds =$$

$$= \int_0^t \operatorname{sp} \mathbf{M}\, [Z_s^*(\omega)\, Z_s(\omega) + Z^*_{\frac{[ns]}{n}}(\omega)\, Z_{\frac{[ns]}{n}}(\omega) - Z_s^*(\omega)\, Z_{\frac{[ns]}{n}}(\omega) -$$

$$- Z^*_{\frac{[ns]}{n}}(\omega)\, Z_s(\omega)]\, R_x ds = \int_0^t \operatorname{sp}\, [B_{ss} + B_{\frac{[ns]}{n}\frac{[ns]}{n}} - B_{\frac{[ns]}{n}s} - B_{s\frac{[ns]}{n}}]\, R_x ds =$$

$$= \int_0^t \sum_{k=1}^\infty \lambda_k(x)\, ([B_{ss} + B_{\frac{[ns]}{n}\frac{[ns]}{n}} - B_{\frac{[ns]}{n}s} - B_{s\frac{[ns]}{n}}]\, f_k(x),\ f_k(x))\, ds.$$

The last expression tends to zero, since the series in the integrand is convergent uniformly in n (because $|f_k| \leqslant 1$, $\|B_{su}\| \leqslant \|B_{tt}\|$ for s, $u \leqslant t$), while the terms of the series tend to zero as $n \to \infty$ (because $(B_{su}f_k, f_k)$ is jointly continuous in its variables).■

Now let $Y_t(\omega)$ be a non-homogeneous continuous Gaussian process with independent increments, with values in $\mathbf{L}_s(\Omega, \mathbf{X})$ (a non-homogeneous Wiener process). Denote

$$\beta_t(x, u, y, v) = \mathbf{M}(Y_t(\omega)x, y)(Y_t(\omega)y, v).$$

Assume that there exists a continuous nondecreasing function $\beta(t)$, such that

$$\beta_t(x, u, x, v) = \int_0^t (R_x(s)u, v)\, d\lambda(s),$$

where $R_x(s)$ is a nuclear operator which is a quadratic function of x and is weakly continuous in s for every x. Then, as before, we can construct an integral $I^*(Z, Y)_t$ for all functions $Z_s(\omega)$ in $\hat{T}_\beta$, where $\hat{T}_\beta$ is the minimal class of functions $Z_t(\omega)$ satisfying conditions (a) and (b) above, and also the following condition (c): if $Z_t^{(n)}(\omega) \in \hat{T}_\beta$ and $Z_t(\omega)$ are such that for all $x, t > 0$,

$$\lim_{n\to\infty} \int_0^t \operatorname{sp}\,[Z_s(\omega) - Z_s^{(n)}(\omega)]\, R_x(s)\, [Z_s(\omega) - Z_s^{(n)}(\omega)]^*\, d\lambda(s) = 0,$$

then $Z_t(\omega) \in \hat{T}_\beta$. The integral $I^*(Z, Y)_t$ is a continuous operator-valued martingale, for whose characteristic

$$(\langle I^*(Z, Y)_t\rangle x, x) = \int_0^t \operatorname{sp} Z_s(\omega)\, R_x(s)\, Z_s^*(\omega)\, d\lambda(s). \qquad (4.49)$$

Lemma 3. Assume that the operator-valued function $Z_s(\omega)$ satisfies condition (a) and is strongly continuous in mean square; let $R_x(s)$ be such that

$$\lim_{s\to t} \operatorname{sp}|R_x(s) - R_x(t)| = 0, \qquad (4.50)$$

for any $t \geqslant 0$ and $x \in \mathbf{X}$ (for a symmetric bounded operator A, $|A|$ is defined as the function $|\cdot|$ of the operator). Then

$Z_s(\omega) \in \hat{T}_\beta$.

Proof. Using the notation of the lemma, we obtain

$$\mathbf{M}\int_0^t \operatorname{sp}[Z_s(\omega) - Z_s^{(n)}(\omega)]\, R_x(s)\, [Z_s(\omega) - Z_s^{(n)}(\omega)]^*\, d\lambda(s) =$$

$$= \int_0^t \operatorname{sp} B_{\frac{[ns]}{n}s} R_x(s)\, d\lambda(s) = \sum_{l=0}^{m-1} \int_{\frac{lt}{m}}^{\frac{l+1}{m}t} \operatorname{sp} B_{\frac{[ns]}{n}s} R_x\left(\frac{lt}{m}\right) d\lambda(s) +$$

$$+ \sum_{l=0}^{m-1} \int_{\frac{lt}{m}}^{\frac{l+1}{m}t} \operatorname{sp} B_{\frac{[ns]}{n}s} \left[R_x(s) - R_x\left(\frac{lt}{m}\right)\right] d\lambda(s),$$

where $[ns]$ is the integral part of ns. The first sum on the right tends to zero for any m (this is proved as in Lemma 2). The second sum is bounded in absolute value by the expression

$$\gamma \sum_{l=0}^{m-1} \int_{\frac{lt}{m}}^{\frac{l+1}{m}t} \operatorname{sp}\left| R_x(s) - R_x\left(\frac{lt}{m}\right)\right| d\lambda(s) \leqslant$$

$$\leqslant \gamma\lambda(t) \sup_{|u-s| \leqslant \frac{t}{m}, u,s \leqslant t} \operatorname{sp}|R_x(s) - R_x(u)|,$$

where $\gamma \geqslant \|B_{su}\|$, and the expression on the right of this inequality tends to zero as $m \to \infty$, by virtue of (4.50). ■

Remark. It is readily seen that $\int_0^t Z_s(\omega) dY_s(\omega)$ is strongly stochastically continuous in mean square, since

$$\mathbf{M}\left|\int_t^{t+h} Z_s(\omega)\, dY_s(\omega)\, x\right|^2 \leqslant \sup_{t \leqslant s \leqslant t+h} \|\mathbf{M}Z_s^*(\omega)\, Z_s(\omega)\| \int_t^{t+h} R_x(s)\, ds.$$

12. Stochastic Operator Equations

12.1. Operator-valued Functions of Random Operators

In our discussion of stochastic operator equations, we shall need operator-valued functions (these are the coefficients of the equations) of the random operators which constitute the range of the unknown operator-valued function. We shall show that, given any operator-valued function $R(A)$ defined on $\mathbf{L}(\mathbf{X})$ and satisfying certain smoothness conditions with respect to A, one can substitute for A a random operator from $\mathbf{L}_s(\Omega, \mathbf{X})$. When this is done, $R(A(\omega))$, where $A(\omega) \in \mathbf{L}_s(\Omega, \mathbf{X})$, is defined via a limit procedure.

Lemma. Let $R(A)$ be defined on $\mathbf{L}(\mathbf{X})$ with values in $\mathbf{L}(\mathbf{X})$, and assume that for all $A, B \in \mathbf{L}(\mathbf{X})$

$$(R(A) - R(B))^* (R(A) - R(B))^* \leqslant \gamma (A - B)^* (A - B).$$

If P_n is the projection operator onto a finite-dimensional subspace, then for all $A(\omega) \in \mathbf{L}_s(\Omega, \mathbf{X})$ and $x \in \mathbf{X}$ the following limit exists, in the sense of convergence in probability:

$$\lim_{n\to\infty} R(A(\omega) P_n) x.$$

Proof. Obviously, $A(\omega) P_n \in \mathbf{L}(\Omega, \mathbf{X})$ with probability 1. Therefore,

$$|R(A(\omega) P_n) x - R(A(\omega) P_m) x|^2 = ([R(A(\omega) P_n) - R(A(\omega) P_m)]^* [R(A(\omega) P_n) - R(A(\omega) P_m)] x, x) \leqslant \gamma |A(\omega)(P_n x - P_m x)|^2.$$

Since $P_n x - P_m x \to 0$ as $n, m \to \infty$ it follows that $|A(\omega)(P_n x - P_m x)| \to 0$ in probability as $A(\omega) \in \mathbf{L}_s(\Omega, \mathbf{X})$. ■

Remark. Let $R(s, A)$ be defined on $[0, \infty) \times \mathbf{L}(\mathbf{X})$ with values in $\mathbf{L}(\mathbf{X})$ satisfying the assumptions of the lemma with γ independent of s, and suppose that for any $A \in \mathbf{L}(\mathbf{X})$ the function R is strongly continuous with respect to s. Then

$$R(s, A(\omega)) = \lim_{n\to\infty} R(s, A(\omega) P_n)$$

is strongly continuous with respect to s, with probability 1. This follows from the fact that $R(s, A(\omega) P_n)$ is strongly continuous and that $R(s, A(\omega) P_n)$ is uniformly convergent in probability; this in turn follows from the fact that the estimate obtained in the lemma is valid uniformly in s.

12.2. Stochastic Equations Involving $I(Y, Z)_t$

Consider the stochastic equation

$$X_t(\omega) = X_0(\omega) + \int_0^t A(s, X_s(\omega))\, ds + \int_0^t dY_s(\omega) B(s, X_s(\omega)), \tag{4.51}$$

where $\{Y_s(\omega), \mathfrak{f}_s\}$ is a continuous strong local martingale, $X_t(\omega)$ an unknown function with values in $\mathbf{L}_s(\Omega, \mathbf{X})$, $X_0(\omega) \in \mathbf{L}_s(\Omega, \mathbf{X})$ an $\mathfrak{f}_0$-measurable random variable, $A(s, Z)$ and $B(s, Z)$ given operator-valued functions on $\mathbf{L}(\mathbf{X}) \times [0, \infty)$ which satisfy the following condition for some $\gamma > 0$ for all Z_1 and $Z_2 \in \mathbf{L}_s(\Omega, \mathbf{X})$;

$$\begin{aligned}&[A(s, Z_1) - A(s, Z_2)]^* [A(s, Z_1) - A(s, Z_2)] + \\ &+ [B(s, Z_1) - B(s, Z_2)]^* [B(s, Z_1) - B(s, Z_2)] \leqslant \\ &\leqslant \gamma (Z_1 - Z_2)^* (Z_1 - Z_2).\end{aligned} \tag{4.52}$$

Then, by the lemma proved above, $A(s, X_s(\omega))$ and $B(s, X_s(\omega))$ are defined for any $\mathfrak{f}_s$-measurable operator $\mathbf{X}_s(\omega)$ and, as is readily deduced from the proof of the lemma, both operators are $\mathfrak{f}_s$-measurable. $\mathbf{X}_t(\omega)$ will be a solution of equation (4.51) if both integrals on the right of (4.51) are defined (the first of them is defined in the strong sense), i.e., for any $x \in \mathbf{X}$ the integral

$$\int_0^t A(s, X_s(\omega))\, x ds$$

exists, and for any t equality holds in (4.51) with probability 1. Since the left-hand side of the equation is stochastically equivalent to a strongly continuous operator process, we shall consider only strongly continuous solutions (the integrals of stochastically equivalent functions are themselves stochastically equivalent).

Theorem 1. Let $A(s, Z)$, $B(s, Z)$ be functions that satisfy condition (4.52) and are strongly continuous in s for fixed $Z \in \mathbf{L}(\mathbf{X})$; let the characteristic of the martingale $\{(Y_s(\omega), \mathfrak{f}_s\}$ admit a representation

$$\langle Y\rangle_t(\omega) = \int_0^t \Phi_s^*(\omega)\Phi_s(\omega)\,ds,$$

where $\Phi_s(\omega) \in \mathbf{L}(\Omega, \mathbf{X})$ is $\mathfrak{f}_s$-measurable and there exists an increasing continuous $\mathfrak{f}_t$-measurable function $\gamma_t(\omega)$, such that $\|\Phi_s(\omega)\| = \|\Phi_s^*(\omega)\| \leqslant \gamma_t(\omega)$ for $s \leqslant t$. Then equation (4.51) has a solution $X_t(\omega)$, which is moreover unique up to stochastic equivalence.

Proof. We first establish uniqueness. Suppose that $X_t(\omega)$ and $X_t'(\omega)$ are strongly continuous solutions (as mentioned above, we are considering only such solutions). Let τ be a Markov time for the filtration $\mathfrak{f}_t$, such that for $s \leqslant \tau$ one has $\|X_s(\omega)x\| + \|X_s'(\omega)x\| + \gamma_s(\omega) \leqslant \alpha$, where $x \in \mathbf{X}$, $\alpha > 0$. If $\alpha \uparrow \infty$, then $\tau \uparrow \infty$. Writing equation (4.51) for $X_s'(\omega)$ and subtracting, we deduce that for any $x \in \mathbf{X}$

$$[X_t(\omega) - X_t'(\omega)]\,x = \int_0^t [A(s, X_s(\omega) - A(s, X_s'(\omega)]\,x ds +$$

$$+ \int_0^t t dY_s(\omega)\,[B(s, X_s(\omega)) - B(s, X_s'(\omega))]\,x. \qquad (4.53)$$

Note that

$$\left|\int_0^t [A(s, X_s(\omega)) - A(s, X_s^*(\omega))]\,x ds\right|^2 \leqslant t\int_0^t |\,[A(s, X_s(\omega)) -$$

$$- A(s, X_s'(\omega))]\,x\,|^2 ds \leqslant \gamma t \int_0^t |\,[X_s(\omega) - X_s'(\omega)]\,x\,|^2 ds.$$

Now, the second term on the right of (4.53) is a continuous local martingale with characteristic (see formula (4.26))

$$\int_0^t |\,\Phi_s(\omega)\,[B(s, X_s(\omega)) - B(s, X_s^*(\omega))]\,x\,|^2 ds \leqslant \int_0^t \gamma_s(\omega)\,\times$$

$$\times\,|\,[B(s, X_s(\omega)) - B(s, X_s'(\omega))]\,x\,|^2 ds \leqslant \gamma_t(\omega)\,\gamma\,\times$$

$$\times \int_0^t |\,[X_s(\omega) - X_s'(\omega)]\,x\,|^2 ds.$$

By our assumptions concerning τ,

$$\gamma_{t\wedge\tau}(\omega)\int\limits_0^{t\wedge\tau}|[X_s(\omega)-X'_s(\omega)]|^2\,ds\leqslant 4\alpha^3 t.$$

Hence one has

$$\mathbf{M}\int\limits_0^{t\wedge\tau}dY_s(\omega)[B(s,\ X_s(\omega))-B(s,\ X'_s(\omega))\,x|^2=$$

$$=\mathbf{M}\int\limits_0^{t\wedge\tau}|\Phi_s(\omega)[B(s,\ X_s(\omega))-B(s,\ X'_s(\omega))]\,x|^2\,ds\leqslant$$

$$\leqslant\gamma\mathbf{M}\int\limits_0^{t\wedge\tau}|[X_s(\omega)-X'_s(\omega)]\,x|^2\,ds,$$

and similarly

$$\mathbf{M}\left|\int\limits_0^{t\wedge\tau}[A(s,\ X_s(\omega))-A(s,\ X'_s(\omega))]\,x\,ds\right|^2\leqslant$$

$$\leqslant\gamma t\mathbf{M}\int\limits_0^{t\wedge\tau}|[X_s(\omega)-X'_s(\omega)]\,x|^2\,ds.$$

Replacing t in (4.53) by $t\wedge\tau$, taking absolute values and squaring, we obtain

$$\mathbf{M}|[X_{t\wedge\tau}(\omega)-X'_{t\wedge\tau}(\omega)]\,x|^2\leqslant 2\mathbf{M}\int\limits_0^{t\wedge\tau}[A(s,\ X_s(\omega))-$$

$$-A(s,\ X'_s(\omega))]\,x\,ds|^2+2\mathbf{M}\int\limits_0^{t\wedge\tau}dY_s(\omega)[B(s,\ X_s(\omega))-$$

$$-B(s,\ X'_s(\omega))]\,x\leqslant 2\gamma(t+\alpha)\mathbf{M}\int\limits_0^{t\wedge\tau}|[X_s(\omega)-X'_s(\omega)]\,x|^2\,ds\leqslant$$

$$\leqslant 2\gamma(t+\alpha)\int\limits_0^t\mathbf{M}|[X_{s\wedge\tau}(\omega)-X'_{s\wedge\tau}(\omega)]\,x|^2\,ds. \qquad (4.54)$$

Denote $\alpha(t)=\mathbf{M}|[X_{t\wedge\tau}(\omega)-X'_{t\wedge\tau}(\omega)]\,x|^2$. Then for any $T>0$ we can find β_T such that $\alpha(T)\leqslant\beta_T\int\limits_0^t\alpha(s)\,ds.$ for $t\leqslant T$. Since

$\alpha(t) \geqslant 0$, it follows that $\alpha(t) = 0$ (see, e.g., [7], p. 41). Thus

$$\mathbf{M}\,|\,[X_{t\wedge\tau}(\omega) - X'_{t\wedge\tau}(\omega)]\,x\,|^2 = 0.$$

Letting $\alpha \uparrow \infty$ (α occurs in the definition of τ), we obtain $\mathbf{M}\,|\,[\mathbf{X}_t(\omega) - \mathbf{X}'_t(\omega)]\,x\,|^2$ for all $t > 0$, $x \in \mathbf{X}$. This proves uniqueness.

We now prove existence of a solution. It follows from the uniqueness proof that stochastic operator equations may be treated in analogy with conventional stochastic differential equations (see, e.g., [7], § 6, Ch. 2). In the existence proof, therefore, we shall exhibit the details of only those parts of the proof that involve arguments specific to the operator case. The solution of equation (4.51) will be constructed by successive approximations. We put

$$X_t^0(\omega) = X(\omega), \quad X_t^n(\omega) = X(\omega) + \int_0^t A(s, X_s^{n-1}(\omega))\,ds +$$

$$+ \int dY_s(\omega)\,B(s, X_s^{n-1}(\omega)), \quad n \geqslant 1. \qquad (4.55)$$

We note that the integrals on the right of (4.55) are defined. Indeed, the function $A(s, X_s^{n-1}(\omega))$ is strongly continuous in s, provided $X_s^{n-1}(\omega)$ is strongly continuous, since

$$|A(s, X_s^{n-1}(\omega))\,x - A(u, X_s^{n-1}(\omega))\,x\,|^2 \leqslant 2\,|\,A(s, X_s^{n-1}(\omega))\,x -$$
$$- x - A(u, X_s^{n-1}(\omega))\,x\,|^2 + 2\,|\,A(s, X_s^{n-1}(\omega))\,x -$$
$$- A(u, X_u^{n-1}(\omega))\,x\,|^2 \leqslant 2\,|\,A(s, X_s^{n-1}(\omega))\,x -$$
$$- A(u, X_s^{n-1}(\omega))\,x\,|^2 + 2\gamma\,|\,X_s^{n-1}(\omega)\,x - X_u^{n-1}(\omega)\,x\,|^2,$$

and both terms on the right tend to zero as $u \to s$ (the first by virtue of the Remark to the Lemma). The stochastic integral in (4.55) also exists (see § 11), since

$$B(s, X_s^{n-1}(\omega))\,x \in \mathcal{F}_{\langle Y \rangle}$$

thanks to the existence of the integral

$$\int_0^t |\Phi_s^*(\omega) B(s, X_s^{n-1}(\omega)) x|^2 ds,$$

provided $X_s^{n-1}(\omega)$ is strongly continuous (in that case $B(s, X_s^{n-1}(\omega))$ is also strongly continuous - the proof is exactly the same as for $A(s, X_s^{n-1}(\omega))$). Thus, if $X_t^{n-1}(\omega)$ is defined and strongly continuous, then $X_t^n(\omega)$ is also defined and consequently strongly continuous. Since $X_t^0(\omega)$ is strongly continuous, it follows by induction that all $X_t^n(\omega)$ are defined and strongly continuous. Consider the difference

$$X_t^{n+1}(\omega) - X_t^n(\omega) = \int_0^t [A(s, X_s^n(\omega)) - A(s, X_s^{n-1}(\omega))] ds +$$

$$+ \int_0^t dY_s(\omega) [B(s, X_s^n(\omega)) - B(s, X_s^{n-1}(\omega))].$$

Let τ be a Markov time such that $\gamma_s(\omega) \leqslant \beta$ for $s \leqslant \tau$. Proceeding as in the case of (4.54), except that the unconditional expectations must be replaced by expectations conditional on the σ-algebra $\mathfrak{f}_0$, we find that for $t \leqslant \tau$

$$\mathbf{M}(|X_{t\wedge\tau}^{n+1}(\omega) x - X_{t\wedge\tau}^n(\omega) x|^2/\mathfrak{f}_0) \leqslant 2\gamma(T + \beta) \times$$

$$\times \int_0^t \mathbf{M}(|X_{s\wedge\tau}^n(\omega) x - X_{s\wedge\tau}^{n-1}(\omega) x|^2/\mathfrak{f}_0) ds. \tag{4.56}$$

To verify that the right-hand side of (4.56) is finite, we show that it is finite for $n = 1$:

$$\int_0^t \mathbf{M}(|X_{s\wedge\tau}^1(\omega) x - X_{s\wedge\tau}^0(\omega) x|^2/\mathfrak{f}_0) ds = \int_0^t \mathbf{M}\left(\left|\int_0^{s\wedge\tau} A(u,\right.\right.$$

$$\left.\left. X_0(\omega)) du + \int_0^{s\wedge\tau} dY_u(\omega) B(u, X_0(\omega))\right) x\right|^2/\mathfrak{f}_0\Big) ds \leqslant$$

$$\leqslant \int_0^t \int_0^{s\wedge\tau} [|A(u, X_0(\omega)) x|^2 + \mathbf{M}\{|\Phi_u(\omega) B(u, X_0(\omega)) x|^2/\mathfrak{f}_0\}] \times$$

$$\times duds \leqslant (t + \beta) \int_0^t \int_0^s (|A(u, X_0(\omega)) x|^2 +$$

$$+ |B(u, X_0(\omega x|^2) duds.$$

It follows from (4.52) and the weak continuity of $A(u, 0)$ and $B(u, 0)$ that for any $T > 0$ there exists β_T, such that

$$|A(u, Z)x|^2 + |B(u, Z)x|^2 \leqslant \gamma_T(|x|^2 + |Zx|^2). \qquad (4.57)$$

Therefore,

$$\int_0^t \mathbf{M}(|X^1_{s\wedge\tau}(\omega)x - X^0_{s\wedge\tau}(\omega)x|^2/\mathfrak{f}_0)\,ds \leqslant \gamma_T(T+\beta)\frac{t^2}{2}(|x|^2 + $$

$$+ |X_0(\omega)x|^2).$$

Setting $\hat{\gamma}_T = \gamma_T \wedge 2\gamma(T+\beta)$, we deduce by induction from (4.56) that

$$\mathbf{M}(|X^{n+1}_{t\wedge\tau}(\omega)x - X^n_{t\wedge\tau}(\omega)x|^2/\mathfrak{f}_0) \leqslant \frac{(\hat{\gamma}_T T)^{n+1}}{(n+1)!}(|x|^2 + |X(\omega)|^2).$$

Using the inequality for the supremum of a martingale, we can write

$$\mathbf{M}(\sup_{t\leqslant T}|X^{n+1}_{t\wedge\tau}(\omega)x - X^n_{t\wedge\tau}(\omega)x|^2/\mathfrak{f}_0) \leqslant$$

$$\leqslant 2T\mathbf{M}\left(\int_0^{T\wedge\tau}|A(s, X^n_s(\omega))x - A(s, X^{n-1}_s(\omega))x|^2\,ds/\mathfrak{f}_0\right) +$$

$$+ 8\mathbf{M}\left(\left|\int_0^{T\wedge\tau} dY_s(\omega)[B(s, X^n_s(\omega))x - B(s, X^{n-1}_s(\omega))x]^2/\mathfrak{f}_0\right) \leqslant\right.$$

$$\leqslant 4\frac{(\hat{\gamma}_T T)^n}{n!}(|x|^2 + |X_0(\omega)x|^2).$$

It follows from this inequality that for any $T > 0$, $x \in \mathbf{X}$, the sequence of $\mathbf{X}$-valued processes $X^n_{t\wedge\tau}(\omega)x$ converges uniformly with probability 1 to some process $X^\tau_t(\omega)$ on [0, T]. Therefore $X^\tau_t(\omega)$, as a strong limit of operators in $\mathbf{L}_s(\Omega, \mathbf{X})$, is also an operator in $\mathbf{L}_s(\Omega, \mathbf{X})$. Using the inequalities

$$\left|\int_0^t A(s, X^n_{s\wedge\tau}(\omega))x\,ds - \int_0^t A(s, X^\tau_s(\omega))x\,ds\right|^2 \leqslant t\gamma\int_0^t |X^n_s(\omega)x -$$

$$- X^\tau_s(\omega)x|^2\,ds,$$

$$(\langle I(Y, B(s, X^n_{s\wedge\tau}) - B(s, X^\tau_s)))_t x, x) \leqslant$$
$$\leqslant \gamma \int_0^t |X^n_{s\wedge\tau}(\omega)x - X^\tau_s(\omega)x|^2 ds$$

and letting $n \to \infty$ in (4.55) with t replaced by $t \wedge \tau$, we obtain

$$X^\tau_t(\omega) = X(\omega) + \int_0^{t\wedge\tau} A(s, X^\tau_s(\omega))\, ds + \int_0^{s\wedge\tau} dY_s(\omega) B(s, X^\tau_s(\omega)).$$

It is also clear that if τ' is another Markov time, appropriate to $\beta' > \beta$, then $X^\tau_s(\omega) = X^{\tau'}_s(\omega)$ for $s \leqslant \tau$. Hence $X^\tau_t(\omega)$ has a limit as $\beta \uparrow \infty$, say $X_t(\omega)$. This limit is a solution of equation (4.51).■

Remark. Let $\gamma_t(\omega) \leqslant \gamma_1$. Then τ may be replaced in (4.56) by $+\infty$. In addition, suppose that $\mathbf{M}|X_0(\omega)x|^2 < \infty$ for all x. Then the expectations in (4.56) may be assumed unconditional; summation over x then yields the inequality

$$\|\mathbf{M}(X^{n+1}_t(\omega) - X^n_t(\omega))^*(X^{n+1}_t(\omega) - X^n_t(\omega))\| \leqslant$$
$$\leqslant 2\gamma(T+\gamma_1)\int_0^t \|\mathbf{M}(X^n_s(\omega) - X^{n-1}_s(\omega))(X^n_s(\omega) - X^{n-1}_s(\omega))\|\, ds.$$

Moreover, the estimates for $n = 0$ give

$$\|\mathbf{M}(X^1_t(\omega) - X^0_t(\omega))^*(X^1_t(\omega) - X^0_t(\omega))\| \leqslant$$
$$\leqslant (t+\gamma_1)\gamma_T \int_0^t (1 + \|\mathbf{M}X^*_0(\omega)X_0(\omega)\|)\, ds.$$

In this case, therefore,

$$\|\mathbf{M}(X^{n+1}_t(\omega) - X^n_t(\omega))^*(X^{n+1}_t(\omega) - X^n_t(\omega))\| = O\left(\frac{t^{n+1}}{(n+1)!}\right);$$
$$\|\mathbf{M}(X_t(\omega) - X^n_t(\omega))^*(X_t(\omega) - X^n_t(\omega)\| \to 0.$$

(This follows from the inequality

$$\|\mathbf{M}(X_t(\omega) - X^n_t(\omega))^*(X_t(\omega) - X^n_t(\omega))\|^{1/2} \leqslant$$
$$\leqslant \sum_{m=n}^{\infty} \|\mathbf{M}(X^{m+1}_t(\omega) - X^m_t(\omega))^*(X^{m+1}_t(\omega) - X^m_t(\omega))\|^{1/2}.)$$

12.3. Stochastic Equations Involving $I^*(Z, Y)t$

We now consider a stochastic equation

$$X_t(\omega) = X_0(\omega) + \int_0^t A(s, X_s(\omega))\,ds + \int_0^t B(s, X_s(\omega))\,dY_s(\omega) \tag{4.58}$$

(notation as in the preceding subsection). We limit ourselves to the case in which $Y_t(\omega)$ is a strongly continuous Gaussian process with independent increments, taking values in $\mathbf{L}_s(\Omega, \mathbf{X})$. Assume that

$$\mathbf{M}(Y_t(\omega)x, u)\,\mathbf{M}(Y_t(\omega)x, v) = \int_0^t (R_x(s)u, v)\,ds, \tag{4.59}$$

where $R_x(s)$ is a nuclear operator, a quadratic function of x, satisfying the condition

$$\lim_{s\to t} \operatorname{sp} |R_x(s) - R_x(t)| = 0. \tag{4.60}$$

As shown in subsection 3, § 11, a sufficient condition for the existence of the stochastic integral in (4.58) in this case is that the function

$$\mathbf{M}\,|B(s, X_s(\omega)x - B(t, X_t(\omega))x|^2$$

tend to zero as $s \to t$ (it will then be continuous in s and t). This last expression is bounded above by

$$2\mathbf{M}\,|B(s, X_t(\omega))x - B(t, X_t(\omega))x|^2 + \\ + 2\mathbf{M}\,|B(s, X_s(\omega))x - B(s, X_t(\omega))x|^2.$$

Assume that (4.52) holds and that $B(s, Z)$ is strongly continuous in s. Then (4.57) will hold and so

$$|B(s, X_t(\omega))x - B(t, X_t(\omega))x|^2 \leqslant 4\gamma((x, x) + |X_t(\omega)x|^2),$$

so that, if $\mathbf{M}\,|X_t(\omega)x|^2 < \infty$, then

$$\lim_{s\to t} \mathbf{M}\,|B(s, X_t(\omega))x - B(t, X_t(\omega))x|^2 = 0$$

by the Lebesgue Theorem, since $|B(s, X_t(\omega))x - B(t, X_t(\omega))x| \to 0$ in probability. For the second term in (4.61) we obtain the estimate

$$\mathbf{M}\,|B(s,\ X_s(\omega))\,x - B(s,\ X_t(\omega))\,x\,|^2 \leqslant$$
$$\leqslant \gamma\mathbf{M}\,|X_s(\omega)\,x - X_t(\omega)\,x\,|^2.$$

The right-hand side tends to zero if $X_s(\omega)\,x$ is continuous in s in mean square. Thus, under the above assumptions on $B(s, Z)$, the stochastic integral $I^*(B(s, X), Y)_t$ is defined if $X_s(\omega)\,x$ is continuous in s in mean square for every x. In that case the integral is strongly continuous in mean square:

$$\mathbf{M}\left|\int_0^t B(s,\ X_s(\omega))\,dY_s(\omega)\,x - \int_0^{t+h} B(s,\ X_s(\omega))\,dY_s(\omega)\,x\right|^2 =$$
$$= \mathbf{M}\left|\int_t^{t+h} B(s,\ X_s(\omega))\,dY_s(\omega)\,x\right|^2 =$$
$$= \int_t^{t+h} \mathbf{M}\,\mathrm{sp}\,B^*(s,\ X_s(\omega))\,B(s,\ X_s(\omega))\,R_x(s)\,ds \leqslant$$
$$\leqslant \gamma_T \int_t^{t+h} \mathbf{M}\,\mathrm{sp}\,(I + X_s^*(\omega)\,X_s(\omega))\,R_x(s)\,ds \qquad (4.62)$$

(it follows from (4.57) that $B^*B \leqslant \gamma_T(I + Z^*Z)$).

Since $\mathbf{M}\,|\,X_s(\omega)\,x\,|^2$ is bounded as a function of s for all x, it follows that $\mathbf{M}\,X_s^*(\omega)X_s(\omega) = V_s$ is a bounded operator-valued function. Therefore, the right-hand side of (4.62) tends to zero uniformly $t \leqslant T$ as $h \to 0$, whatever T.

Using (4.57), (4.52), we can also prove that the integral

$$\int_0^t A(s,\ X_s(\omega))\,ds$$

exists and is strongly continuous, provided the process $X_s(\omega)$ is strongly continuous in mean square:

$$\mathbf{M}\left|\int_0^{t+h} A(s,\ X_s(\omega))\,dsx - \int_0^t A(s,\ X_s(\omega))\,dsx\right|^2 =$$
$$= \mathbf{M}\left|\int_t^{t+h} A(s,\ X_s(\omega))\,ds\right|^2 \leqslant h\int_t^{t+h} \mathbf{M}\,|A(s,\ X_s(\omega))\,x\,|^2\,ds \leqslant$$
$$\leqslant h\gamma_T \int_t^{t+h} ((x,\ x) + \mathbf{M}\,|\,X_s(\omega)\,x\,|)^2\,ds \quad \text{as} \quad h \to 0.$$

<u>Theorem 2</u>. Let $A(s, Z)$, $B(s, Z)$ be functions satisfying

the assumptions of Theorem 1, and $Y_t(\omega)$ a Gaussian process with independent increments, for which (4.59) and (4.60) hold; let $X_0(\omega)$ be an $\mathfrak{f}_0$-measurable r.v. with values in $\mathbf{L}_s(\Omega, X)$ such that $\mathbf{M}\,|X_0(\omega)x|^2 < \infty$ for all $x \in \mathbf{X}$. Then equation (4.58) has a solution satisfying the following conditions:

(1) $X_t(\omega) \in \mathbf{L}_s(\Omega, \mathbf{X})$ and $X_t(\omega)$ are continuous in t for all x;

(2) $X_t(\omega)$ is $\mathfrak{f}_t$-measurable;

(3) $\lim\limits_{s\to t} \mathbf{M}\,|X_t(\omega)x - X_s(\omega)x|^2 = 0$ for all $x \in \mathbf{X}$.

The solution satisfying these conditions is unique.

Proof. Uniqueness. Let $X_t(\omega)$ and $X'_t(\omega)$ be solutions satsifying the conditions of the theorem. Then, proceeding as in the proof of Theorem 1, we obtain

$$\mathbf{M}\,|X_t(\omega)x - X'_t(\omega)x|^2 \leqslant 2t\gamma \int_0^t \mathbf{M}\,|X_s(\omega)x - X'_s(\omega)x|^2\, ds +$$

$$+ \gamma \mathbf{M} \int_0^t \operatorname{sp}(X_s(\omega) - X'_s(\omega))\, R_x(s)\, (X_s(\omega) - X'_s(\omega))^*\, ds \leqslant$$

$$\leqslant 2t\gamma \int_0^t \|\mathbf{M}\,[X_s(\omega) - X'_s(\omega)]^*\,[X_s(\omega) - X'_s(\omega)]\|\, ds\; x|^2 +$$

$$+ 2\gamma \int_0^t \|\mathbf{M}\,[X_s(\omega) - X'_s(\omega)]^*\,[X_s(\omega) - X'_s(\omega)]\| \operatorname{sp} R_x(s)\, ds. \tag{4.63}$$

By assumption, $R_x(s) = (R(s)x, x)$, where, by (4.60), $R(s)$ is weakly continuous in s and is therefore a bounded function. Let $\|R(s)\| \leqslant \gamma_T$ for $s \leqslant T$ (we shall assume that this is the same constant γ_T as in (4.57)). Then, evaluating the supremum in (4.63) for $|x| \leqslant 1$, we see that

$$\|\mathbf{M}\,(X_t(\omega) - X'_t(\omega))^*\,(X_t(\omega) - X'_t(\omega))\| \leqslant 2\,[T\gamma +$$

$$+ \gamma\gamma_T] \int_0^t \|\mathbf{M}\,(X_s(\omega) - X'_s(\omega))^*\,(X_s(\omega) - X'_s(\omega))\|\, ds, \tag{4.64}$$

whence it follows that

$$\|\mathbf{M}\,(X_t(\omega) - X'_t(\omega))^*\,(X_t(\omega) - X'_t(\omega))\| = 0.$$

Therefore

$$(\mathbf{M}(X_t(\omega) - X_t'(\omega))^*(X_t(\omega) - X_t'(\omega))x, \ x) = \mathbf{M}\,|X_t(\omega)x - X_t'(\omega)x|^2 = 0$$

for all $x \in \mathbf{X}$. This proves uniqueness.

The existence of the solution is again proved by the successive approximation method. Put

$$X_t^0(\omega) = X_0(\omega), \ X_t^{n+1}(\omega) = X_0(\omega) + \int_0^t A(s, \ X_s^n(\omega))\,ds + \\ + \int_0^t B(s, \ X_s^n(\omega))\,dY_s(\omega). \qquad (4.65)$$

If $X_s^n(\omega)$ is strongly continuous in mean square, the existence of the integrals on the right has aleady been established, and moreover both integrals are then strongly continuous in mean square. Since $X_s^0(\omega)$ is strongly continuous in mean square, induction establishes the existence of all integrals on the right of (4.65) and the strong continuity in mean square of $X_t^n(\omega)$ for all n.

Substracting equalities (4.65) for two consecutive n--values from one another, and using the same estimates as in (4.63), (4.64), we obtain $(t \leqslant T)$

$$\|\mathbf{M}(X_t^{n+1}(\omega) - X_t^n(\omega))^*(X_t^{n+1}(\omega) - X_t^n(\omega))\| \leqslant \\ \leqslant 2\gamma(T + \gamma_T)\int_0^t \|\mathbf{M}(X_s^n(\omega) - X_s^{n-1}(\omega))^*(X_s^n(\omega) - \\ - X_s^{n-1}(\omega))\|\,ds. \qquad (4.66)$$

Putting $2\gamma\,[T + \gamma_T] = \beta_T$, we obtain from (4.66)

$$\|\mathbf{M}(X_t^{n+1}(\omega) - X_t^n(\omega))^*(X_t^{n+1}(\omega) - X_t^n(\omega))\| \leqslant \alpha_T \frac{(\beta_T T)^{n+1}}{(n+1)!};$$

$$\alpha_T = \frac{\gamma_T}{\gamma}(1 + \|\mathbf{M}X_0^*(\omega)X_0(\omega)\|). \qquad (4.67)$$

Indeed, by (4.57),

$$\|\mathbf{M}(X_t^1(\omega) - X_t^0(\omega))^*(X_t^1(\omega) - X_t^0(\omega))\| \leqslant 2\|\mathbf{M}\left(\int_0^t A(s, X_0(\omega)) \times\right.$$

$$\left.\times ds\right)^*\left(\int_0^t A(s, X_0(\omega))\, ds\right)\| + 2\|\mathbf{M}\left(\int_0^t B(s, X_0(\omega))\, dY_s(\omega)\right)^* \times$$

$$\times\left(\int_0^t B(s, X_0(\omega))\, ds\right)\| \leqslant [2T \int_0^t \|\mathbf{M}A^*(s, X_0(\omega))\, A(s, X_0(\omega))\| \times$$

$$\times ds + 2\int_0^t \|B^*(s, X_0(\omega))\, B(s, X_0(\omega))\| \sup_{|x| \leqslant 1} \operatorname{sp} R_x(s) \leqslant$$

$$\leqslant (2T + \gamma_T)\gamma_T \int_0^t (1 + \|\mathbf{M}X_0^*(\omega)\, X_0(\omega)\|\, ds =$$

$$= \beta_T \frac{\gamma_T}{\gamma}(1 + \|\mathbf{M}X_0^*(\omega)\, X_0(\omega)\|)\, t = \beta_T\alpha_T t$$

and then (4.67) follows from (4.66) by induction. It follows from the estimate (4.67) that for all $x \in \mathbf{X}$ and $t > 0$ we have a strong limit in probability

$$\lim_{n\to\infty} X_t^n(\omega)\, x = X_t(\omega)\, x,$$

where $X_t(\omega) \in \mathbf{L}_s(\Omega, \mathbf{X})$. And then, for $t \leqslant T$,

$$\sqrt{\mathbf{M}|X_t^n(\omega)\, x - X_t(\omega)\, x|^2} \leqslant$$

$$\leqslant \sum_{m=n}^{\infty} \sqrt{\mathbf{M}|X_t^m(\omega)\, x - X_t^{m+1}(\omega)\, x|^2} \leqslant |x| \sum_{m=n}^{\infty} \|\mathbf{M}(X_t^m(\omega) -$$

$$- X_t^{m+1}(\omega))^*(X_t^m(\omega) - X_t^{m+1}(\omega))\|^{1/2} \leqslant |x| \sum_{m=n}^{\infty} \sqrt{\frac{\alpha_T(\beta_T T)^{m+1}}{(m+1)!}}.$$

Therefore $X_t^n(\omega)\, x$ converges to $X_t(\omega)\, x$ uniformly in mean square. Hence it readily follows that

$$\left\|\mathbf{M}\left(\int_0^t A(s, X_s^n(\omega))\, ds - \int_0^t A(s, X_s(\omega))\, ds\right)^* \times\right.$$

$$\left.\times\left(\int_0^t A(s, X_s^n(\omega))\, ds - \int_0^t A(s, X_s^n(\omega))\, ds\right)\right\| \leqslant$$

$$\leqslant T_\gamma \int_0^T \| \mathbf{M} (X_s^n(\omega) - X_s(\omega))^* (X_s^n(\omega) - X_s(\omega)) \| \, ds \to 0$$

and

$$\left\| \mathbf{M} \left(\int_0^t B(s, X_s^n(\omega)) \, dY_s(\omega) - \int_0^t B(s, X_s(\omega)) \, dY_s(\omega) \right)^* \times \right.$$

$$\left. \times \left(\int_0^t B(s, X_s^n(\omega)) \, dY_s(\omega) - \int_0^t B(s, X_s(\omega)) \, dY_s(\omega) \right) \right\| \leqslant$$

$$\leqslant \int_0^t \| \mathbf{M} (X_s^n(\omega) - X_s(\omega))^* (X_s^n(\omega) - X_s(\omega)) \| \sup_{|x| \leqslant 1} \operatorname{sp} R_x(s) \, ds \to 0.$$

Letting $n \to \infty$ in (4.65), we see that $X_t(\omega)$ is a solution of equation (4.58). That this solution is $\mathfrak{f}_t$-measurable and continuous in mean square follows from the fact that $X_t^n(\omega)$ converges uniformly in mean square to $X_t(\omega)$; that $X_t(\omega)$ is continuous with probability 1 follows from the continuity of the right-hand side of equation (4.58). ■

12.4. Some Generalizations

In this subsection we shall consider equations of type (4.51) and (4.58) whose coefficients satisfy the following condition, which is more general than condition (4.52):

Condition A. There exist a sequence of unitary operators U_k and numbers $\gamma_k \geqslant 0$, such that $\Sigma \gamma_k = \gamma < \infty$ and, for all $Z_1, Z_2 \in \mathbf{L}(\mathbf{X})$, $s \geqslant 0$,

$$[A(s, Z_1) - A(s, Z_2)]^* [A(s, Z_1) - A(s, Z_2)] +$$
$$+ [B(s, Z_1) - B(s, Z_2)]^* [B(s, Z_1) - B(s, Z_2)] \leqslant$$
$$\leqslant \sum_{k=1}^{\infty} \gamma_k U_k^* (Z_1 - Z_2)^* (Z_1 - Z_2) U_k$$

and for any $Z \in \mathbf{L}(\mathbf{X})$ the functions $A(s, Z)$ and $B(s, Z)$ are strongly continuous.

Theorem 3. Suppose that the coefficients of the equation satisfy condition A; let $Y_t(\omega)$ be a strongly continuous Gaussian process with independent increments, satisfying condition (4.59), $X_0(\omega)$ an $\mathfrak{f}_0$-measurable function in $\mathbf{L}_s(\Omega, \mathbf{X})$ such that $\mathbf{M} | X_0(\omega) x |^2 < \infty$ for all $x \in \mathbf{X}$. Then equations (4.51), (4.58) have a (unique) solution which is continuous,

$\mathfrak{f}_t$-measurable and strongly continuous in mean square.

Proof. We shall show that when condition A holds the right--hand sides of equations (4.51), (4.58) are defined for any process $X_t(\omega)$ which is strongly continuous in mean square, and that they are also strongly continuous in mean square. To this end, we note that condition A implies the existence for every T of a constant γ_T, such that

$$|A(s, Z)x|^2 + |B(s, Z)x|^2 \leqslant \gamma_T\left(|x|^2 + \sum_{k=1}^{\infty}\gamma_k|ZU_kx|^2\right). \tag{4.68}$$

This is proved by the same arguments as were used to deduce (4.57) from (4.52). Using this inequality, one readily proves the existence and strong continuity of the integral

$$\int_0^t A(s, X_s(\omega))\,ds.$$

The existence of the integral

$$\int_0^t dY_s(\omega)\,B(s, X_s(\omega))\,x$$

follows from the fact that, for all x, $t \leqslant T$,

$$\mathbf{M}\int_0^t (B^*(s, X_s(\omega))\,R(s)\,B(s, X_s(\omega))\,x,\ x)\,ds \leqslant$$

$$\leqslant \mathbf{M}\int_0^t \|R(s)\|\,|B(s, X_s(\omega))\,x|^2\,ds \leqslant \gamma_T\int_0^t \|R(s)\| \times$$

$$\times\left(|x|^2 + \sum_{k=1}^{\infty}\gamma_k\mathbf{M}\,|X_s(\omega)\,U_kx|^2\right)ds \leqslant \gamma_T|x|^2 \times$$

$$\times\int_0^t \|R(s)\|\,(1 + \gamma\|\mathbf{M}X_s^*(\omega)\,X_s(\omega)\|)\,ds.$$

Similarly, one proves the inequality $(t < t+h < T)$

$$\mathbf{M}\left|\int_t^{t+h} dY_s(\omega)\,B(s, X_s(\omega))\,x\right|^2 \leqslant \gamma_T|x|^2\int_t^{t+h}\|R(s)\|\,(1 + \gamma\times$$

$$\times\|\mathbf{M}X_s^*(\omega)\,X_s(\omega)\|)\,ds,$$

which implies that the integral is strongly continuous in

mean square.

The integral $\int_0^t B(s, X_s(\omega))(dY_s(\omega)$ exists by virtue of the strong continuity in mean square of the function $B(s, X_s(\omega))$ (see subsection 3, § 11). Indeed, let $s \to t$. Then

$$\mathbf{M}\,|B(s, X_s(\omega))\,x - B(t, X_t(\omega))\,x\,|^2 \leqslant 2\mathbf{M}\,|B(s, X_s(\omega))\,x - B(s, X_t(\omega))\,x\,|^2 + 2\mathbf{M}\,|B(s, X_t(\omega))\,x - B(t, X_t(\omega))\,x\,|^2.$$

For the first term we have the inequality

$$\mathbf{M}\,|B(s, X_s(\omega))\,x - B(s, X_t(\omega))\,x\,|^2 \leqslant \leqslant \gamma_T \sum_{k=1}^{\infty} \mathbf{M}\,|X_s(\omega)\,U_k x - X_t(\omega)\,U_k\,x\,|^2,$$

while the second term tends to zero since the expression after the expectation symbol $\mathbf{M}$ tends to zero in probability and is bounded by

$$4\gamma_T\left(|\,x\,|^2 + \sum_{k=1}^{\infty} \gamma_k\,|X_t(\omega)\,U_k x\,|^2\right).$$

As shown in subsection 3, § 11, this stochastic integral is also strongly continuous in mean square.

For the rest of the proof we need certain estimates for the difference of integrals. To this end we use the fact that $|\,U_k\,x\,|^2 = |\,x\,|^2$:

$$\mathbf{M}\left|\int_0^t [A(s, X_s(\omega)) - A(s, X'_s(\omega))]\,x ds\right|^2 \leqslant t\int_0^t \mathbf{M}\,|A(s, X_s(\omega))\,x - A(s, X'_s(\omega))\,x\,|^2\,ds \leqslant t\int_0^t \sum_k \gamma_k \mathbf{M}\,|X_s(\omega)\,U_k x - X'_s(\omega)\,U_k x\,|^2\,ds \leqslant t\gamma \int_0^t \|\mathbf{M}\,(X_s(\omega) - X'_s(\omega))^* \times \times (X_s(\omega) - X'_s(\omega))\,\|\,ds\,|\,x\,|^2. \tag{4.69}$$

Furthermore, $\mathbf{M}\left|\int_0^t dY_s(\omega)\,[B(s, X_s(\omega)) - B(s, X'_s(\omega))]\,x\,\right|^2 \leqslant$

$$\leqslant \int_0^t \|\,R_s\,\| \sum_{k=1}^{\infty} \gamma_k\,|\,\mathbf{M}\,(X_s(\omega) - X'_s(\omega))\,U_k x\,|^2\,ds \leqslant$$

$$\leqslant \gamma \sup_{s\leqslant t} \| R_s \| \int_0^t \mathbf{M} (X_s(\omega) - X_s'(\omega))^* (X_s(\omega) - X_s'(\omega)) \| ds, \qquad (4.70)$$

$$\mathbf{M} \left| \int_0^t [B(s, X_s(\omega)) - B(s, X_s'(\omega))] \, dY_s(\omega) \, x \right|^2 \leqslant$$

$$\leqslant \int_0^t \mathbf{M} \, \mathrm{sp} \, [B(s, X_s(\omega)) - B(s, X_s'(\omega))]^* [B(s, X_s(\omega)) -$$

$$- B(s, X_s'(\omega))] R_x(s) \, ds \leqslant \int_0^t \mathbf{M} \, \mathrm{sp} \sum_{k=1}^{\infty} \gamma_k U_k^* (X_s(\omega) - X_s'(\omega))^* \times$$

$$\times (X_s(\omega) - X_s'(\omega)) U_k R_x(s) \, ds = \int_0^t \mathrm{sp} \sum_{k=1}^{\infty} \gamma_k U_k^* \mathbf{M} (X_s(\omega) -$$

$$- X_s'(\omega))^* (X_s(\omega) - X_s'(\omega)) U_k R_x(s) \, ds \leqslant \gamma \sup_{s\leqslant t} R(s) |x|^2 \times$$

$$\times \int_0^t \| \mathbf{M} (X_s(\omega)) - X_s'(\omega))^* (X_s(\omega) - X_s'(\omega)) \| ds. \qquad (4.71)$$

Now, using inequalities (4.69)-(4.71), we can proceed along lines similar to those followed in Theorems 1 and 2, to prove the uniqueness and existence of the solution (the latter - again by successive approximations).■

CHAPTER 5

LINEAR STOCHASTIC OPERATOR EQUATIONS

13. Generalization of the Stochastic Operator Integral

13.1. General Form of the Linear Equation

Proceeding from the concept of stochastic equations considered in the last section, one should understand the term "linear equation" in the sense of an equation

$$dX_t(\omega) = A(t, X_t(\omega))\,dt + dY_t(\omega)\,B(t, X_t(\omega)) \tag{5.1}$$

or

$$dX_t(\omega) = A(t, X_t(\omega))\,dt + B(t, X_t(\omega))\,dY_t(\omega), \tag{5.2}$$

where $A(t, Z)$ and $B(t, Z)$ are linear functions of $Z \in \mathbf{L}(\mathbf{X})$ with values in $\mathbf{L}(\mathbf{X})$. For the right-hand sides of these equations to be meaningful, one must impart some sense to their coefficients in case the argument $X_t(\omega)$ is not in $\mathbf{L}(\Omega, \mathbf{X})$ but in $\mathbf{L}_s(\Omega, \mathbf{X})$. Let $R(Z)$ be some linear function from $\mathbf{L}(\mathbf{X})$ to $\mathbf{L}(\mathbf{X})$. We shall study the conditions under which $R(A, (\omega)), A(\omega) \in \mathbf{L}_s(\Omega, \mathbf{X})$, is defined.

Lemma. Let the following conditions hold:

(1) $R(Z)$ is continuous on $\mathbf{L}(\mathbf{X})$ in the operator norm, and there exists a basis $\{e_k\}$, such that, for all $x \in \mathbf{X}, Z \in \mathbf{L}(\mathbf{X})$,

$$\lim_{n\to\infty} R(ZP_n)\,x = R(Z)\,x,$$

where P_n is the projection operator onto the subspace spanned by $e_1, \ldots, e_n$;

(2) for all $x, y, u, v \in \mathbf{X}$, the expectation

$$\beta_A(x, y, u, v) = \mathbf{M}\,(A(\omega)\,x, y)\,(A(\omega)\,u, v)$$

(3) for all $x \in \mathbf{X}$ the following series is convergent:

$$\sum_{i,j,k,l} (R(\langle e_i \circ e_j\rangle)\,x,\ R(\langle e_k \circ e_l\rangle)\,\beta_A(e_i, e_j, e_k, e_l). \tag{5.3}$$

Then the sequence of random operators $R(A(\omega)P_n)$ is strongly convergent to some random operator $R(A(\omega)) \in \mathbf{L}_s(\Omega, \mathbf{X})$. Moreover, $\mathbf{M}|R(A(\omega))x|^2$ exists and is equal to (5.3).

Proof. Since

$$A(\omega) P_n = \sum_{k=1}^{n} \langle e_k \circ A(\omega) e_k \rangle \in \mathbf{L}(\Omega, \mathbf{X}),$$

it follows that

$$R(A(\omega) P_n) = \sum_{k=1}^{n} R(\langle e_k \circ A(\omega) e_k \rangle) x =$$

$$= \sum_{k=1}^{n} \sum_{i=1}^{\infty} (A(\omega) e_k, e_i) R(\langle e_k \circ e_i \rangle) x.$$

Therefore,

$$|R(A(\omega) P_n) x|^2 = \left| \sum_{k=1}^{n} \sum_{i=1}^{\infty} (A(\omega) e_k, e_i) R(\langle e_k \circ e_i \rangle) x \right|^2.$$

Since the operator $A(\omega)P_n$ is degenerate, it follows that, with probability 1,

$$\lim_{m\to\infty} \| P_m A(\omega) P_n - A(\omega) P_n \| = 0,$$

and so

$$\lim_{m\to\infty} \| R(P_m A(\omega) P_n) - R(A(\omega) P_n) \| = 0.$$

Therefore,

$$|R(A(\omega) P_n) x|^2 = \lim_{m\to\infty} \left| \sum_{k=1}^{n} \sum_{l=1}^{\infty} (P_m A(\omega) e_k, e_l) R(\langle e_k \circ e_l \rangle) x \right|^2 =$$

$$= \lim_{m\to\infty} \left| \sum_{k=1}^{n} \sum_{i=1}^{m} (A(\omega) e_k, e_i) R(\langle e_k \circ e_i \rangle) x \right|^2 .$$

Consequently, by the Fatou Theorem,

$$\mathbf{M} | R(A(\omega) P_n) x |^2 \leqslant \underline{\lim}_{m\to\infty} \mathbf{M} \left| \sum_{k=1}^{n} \sum_{i=1}^{m} (A(\omega) e_k, e_i) \times \right.$$

$$\left. \times R(\langle e_k \circ e_i \rangle) x \right|^2 = \lim_{m\to\infty} \sum_{k,l=1}^{n} \sum_{i,j=1}^{m} (R(\langle e_k \circ e_i \rangle x, R(\langle e_l \circ e_j \rangle x) \times$$

$$\times \mathbf{M}(A(\omega)e_k,\ e_i)(A(\omega)e_l,\ e_j) = \sum_{k,l=1}^{n}\sum_{i,j=1}^{\infty}(R(\langle e_k \circ e_i\rangle)x,$$

$$R(\langle e_l \circ e_j\rangle)x)\,\beta_A(e_k,\ e_i,\ e_l,\ e_j). \tag{5.4}$$

One shows similarly that for $n < m$,

$$\mathbf{M}\,|R(A(\omega)P_n)x - R(A(\omega)P_m)x|^2 \leqslant$$

$$\leqslant \sum_{n<k,l\leqslant m}\sum_{i,j=1}^{\infty}(R(\langle e_k \circ e_i\rangle)x,\ R(\langle e_l \circ e_j\rangle)x)\,\beta_A(e_k,\ e_i,\ e_l,\ e_j).$$

Hence, by the convergence of the series (5.3),

$$\lim_{n,m\to\infty}\mathbf{M}\,|R(A(\omega)P_n)x - R(A(\omega)P_m)x|^2 = 0,$$

so that $R(A(\omega)P_n)x$ converges in mean square (hence also strongly) to some limit, which we denote (tentatively) by $R(A(\omega))x$. Moreover,

$$\lim_{n\to\infty}\mathbf{M}\,|R(A(\omega)P_n)x|^2 = \mathbf{M}\,|R(A(\omega))x|^2. \tag{5.5}$$

Similar reasoning shows that $R(P_m A(\omega)P_n)x$ converges in mean square as $m \to \infty$ (for fixed n) to $R(A(\omega)P_n)x$, so that

$$\mathbf{M}\,|R(A(\omega)P_n)x|^2 = \lim_{m\to\infty}\mathbf{M}\,|R(P_m A(\omega)P_n)x|^2.$$

Since we now have equality instead of inequality throughout (5.4), it follows, in view of (5.5), that

$$\mathbf{M}\,|R(A(\omega))x| =$$

$$= \lim_{n\to\infty}\sum_{k,l\leqslant n}\sum_{i,j=1}^{\infty}(R(\langle e_k \circ e_i\rangle)x,\ R(\langle e_l \circ e_j\rangle)x)\,\beta_A(e_k,\ e_i,\ e_l,\ e_j).$$

Remark. Since it is the limit of nonnegative bounded quadratic forms (as functions of x), $\mathbf{M}\,|R(A(\omega))x|^2$ is also such a form.

An immediate generalization of both equations (5.1) and (5.2) is the linear equation

$$dX_t(\omega) = A(t,\ X_t(\omega))\,dt + dY_t(\omega)\,B_1(t,\ X_t(\omega)) +$$

$$+ B_2(t,\ X_t(\omega))\,dY_t(\omega),$$

where the coefficients are linear functions of X. Note that the martingale part of the equation (i.e., the part involving the differential $dY_t(\omega)$) of a martingale) is a linear operator--valued function of $X_t(\omega)$ and $dY_t(\omega)$. This observation suggests the general form of the linear operator equation:

$$dX_t(\omega) = A(t, X_t(\omega))\,dt + B(t, X_t(\omega),\ dY_t(\omega)), \tag{5.6}$$

where $B(t, Z, W)$ is a function defined for any t and mapping $\mathbf{L}(\mathbf{X}) \times \mathbf{L}(\mathbf{X})$ into $\mathbf{L}(\mathbf{X})$, linearly in each argument. For equation (5.6) to be meaningful, we must define the stochastic integral

$$\int_0^t B(s, Z_s(\omega),\ dY_s(\omega)), \tag{5.7}$$

and this we now proceed to do.

13.2. A Generalization of the Stochastic Integral

Let $Y_t(\omega)$ be a Gaussian operator-valued process with independent increments, with values in $\mathbf{L}_s(\Omega, \mathbf{X})$, such that

$$\mathbf{M}(Y_t(\omega)x, y) = 0, \quad x, y \in \mathbf{X};$$

$$\mathbf{M}(Y_t(\omega)x, y)(Y_t(\omega)u, v) = \beta_t(x, y, u, v), \ (x, y, u, v \in \mathbf{X}).$$

Assume that

$$\beta_t(x, y, u, v) = \int_0^t \beta^s(x, y, u, v)\,ds,$$

where $\beta^s(x, y, u, v)$ is a locally integrable function of s for all x, y, u, v, and for any $s \geqslant 0$ is a bounded quadrilinear form on $\mathbf{X}^4$. With these assumptions on $Y_t(\omega)$ we shall define the integral (5.7).

Let $B(s, Z_1, Z)$ satsify the following condition for all $s > 0$, $Z_1 \in \mathbf{L}(\mathbf{X})$:

I. $B(s, Z_1, Z)$ is a continuous function of Z in the operator norm, and for some basis $\{e_k\}$ (independent of s) and any $x \in \mathbf{X}$,

$$\lim_{n\to\infty} B(s, Z_1, ZP_n)x = B(s, Z_1, Z)x$$

and the series

$$\sum_{k,i,l,j} (B(s, Z_1, \langle e_k \circ e_i \rangle) x,\ B(s, Z_1, \langle e_l \circ e_j \rangle) x) \beta(e_k, e_i, e_l, e_j) \tag{5.8}$$

is convergent uniformly in s on every finite closed interval. Thanks to the Lemma, this series may be expressed as

$$\mathbf{M} | B(s, Z_1, Y^s(\omega)) x |^2, \tag{5.9}$$

where $Y^s(\omega)$ is a Gaussian random operator in $\mathbf{L}_s(\Omega, \mathbf{X})$, such that $\mathbf{M}Y^s(\omega) = 0$ and

$$\mathbf{M}(Y^s(\omega) x, y)(Y^s(\omega) u, v) = \beta^s(x, y, u, v).$$

By the remark to the Lemma, (5.9) is a bounded quadratic form in x, and so the same is true of

$$\mathbf{M}(B(s, Z_1, Y^s(\omega)) x,\ B(s, Z_2, Y^s(\omega)) x), \quad Z_1, Z_2 \in \mathbf{L}(\mathbf{X}).$$

Thus, we can define an operator-valued function

$$C^\beta(s, Z_1, Z_2) = \mathbf{M}B^*(s, Z_1, Y^s(\omega)) B(s, Z_1, Y^s(\omega)),$$

whose values are operators in $\mathbf{L}(\mathbf{X})$. It is easy to see that it is linear in Z_1 and Z_2, and that

$$C^\beta(s, Z_2, Z_1) = [C^\beta(s, Z_1, Z_2)]^*.$$

Note that for any x

$$(C^\beta(s, Z, Z) x, x)$$

is a nonnegative quadratic real-valued function of Z, since this is true of any function $| R(Z) x |^2$, with R a linear function from $\mathbf{L}(\mathbf{X})$ to $\mathbf{L}(\mathbf{X})$.

In the sequel we shall assume that $C^\beta(s, Z, Z)$ satisfies the following conditions:

II. (a) for $s \geqslant 0$, $C^\beta(s, Z, Z)$ is continuous in Z in the operator norm;

(b) for $Z \in \mathbf{L}(\mathbf{X})$, $C^\beta(s, Z, Z)$ is a strongly continuous function of s;

(c) if $\{e_k\}$ is the basis whose existence is stipulated in condition I, and P_n is projection onto the subspace spanned by $\{e_k,\ k = 1, \ldots, n\}$, then for $s \geqslant 0$, $x \in \mathbf{X}$ and $Z \in \mathbf{L}(\mathbf{X})$,

$$\lim_{n,m \to \infty} (C^\beta(s, P_n Z, P_m Z) x, x) = (C^\beta(s, Z, Z) x, x).$$

Repeating the arguments used to prove the Lemma (with obvious modifications), we see that if $Z_s(\omega) \in \mathbf{L}_s(\Omega, \mathbf{X})$, is such that for any $x \in \mathbf{X}$ the series

$$\sum_{k,l,\bar{l},j} (C^\beta(s, \langle e_k \circ e_i\rangle, \langle e_l \circ e_j\rangle)x, x)\mathbf{M}(Z_s(\omega)e_k, e_l)(Z_s(\omega)e_i, e_j) \tag{5.10}$$

is convergent, then the following limit exists (in the sense of convergence in mean):

$$\lim_{\substack{n\to\infty \\ m\to\infty}} C^\beta(s, Z_s(\omega)P_n, Z_s(\omega)P_m),$$

and we denote the limit by $C^\beta(s, Z_s(\omega), Z_s(\omega))$; under these conditions,

$$\mathbf{M}(C^\beta(s, Z_s(\omega), Z_s(\omega))x, x)$$

is identical with (5.10). Thus, $C^\beta(s, Z_s(\omega), Z_s(\omega))$ is defined as a weak nonnegative operator with finite expectation.

Let $\mathfrak{f}_s$ be a filtration, relative to which $Y_s(\omega)$ is a martingale. Let T^2_β denote the set of operator-valued functions $Z_t(\omega)$ with values in $\mathbf{L}_s(\Omega, \mathbf{X})$, which satisfy the following conditions:

(a) $Z_t(\omega)$ is defined and measurable for $t \geqslant 0$;

(b) $\mathbf{M}|Z_t(\omega)x|^2 < \infty$ for all $x \in \mathbf{X}$;

(c) for all $s > 0$ $C^\beta(s, Z_s(\omega)\ \ Z_s(\omega))$ is defined and $\mathbf{M}(C^\beta(s, Z_s(\omega), Z_s(\omega))x, x) < \infty$ (i.e., the series (5.10) is convergent) and for all $t > 0$, $x \in \mathbf{X}$,

$$\int_0^t \mathbf{M}(C^\beta(s, Z_s(\omega), Z_s(\omega))x, x)\,ds < \infty.$$

We claim that for all $Z_s(\omega) \in T^2_\beta$ the integral (5.7) can be defined in such a way that the following conditions hold:

A. As a function of t, the integral is a strong operator--valued martingale, its operator characteristic having the form

$$\int_0^t C^\beta(s, Z_s(\omega), Z_s(\omega))\,ds.$$

B. The integral is a linear function of $Z_s(\omega)$.

We first consider the integral

$$\int_0^t B(s, Z, dY_s(\omega)),$$

where $Z \in \mathbf{L}(\mathbf{X})$ is independent of both s and ω. We define it as follows:

$$\int_0^t B(s, Z, dY_s(\omega)) = \sum_{k=1}^{\infty} \int_0^t B(s, Z, d\langle e_k \circ Y_s(\omega) e_k\rangle) =$$

$$= \sum_{k=1}^{\infty} \sum_{i=1}^{\infty} \int_0^t B(s, Z, \langle e_k \circ e_i\rangle) \, d(Y_s(\omega) e_k, e_i). \qquad (5.11)$$

The terms on the right of (5.11) are ordinary stochastic integrals of nonrandom operator-valued functions with respect to (mutually dependent) one-dimensional Gaussian processes with jointly independent increments. The joint characteristic of the martingales $(Y_s(\omega) e_k, e_i)$ and $(Y_s(\omega) e_l, e_j)$ is $\beta_s(e_k, e_i, e_l, e_j)$. Hence the characteristic of the martingale

$$\sum_{k=1}^{n} \sum_{i=1}^{m} \int_0^t B(s, Z\langle e_k \circ e_i\rangle) \, d(Y_s(\omega) e_k, e_i)$$

will be

$$\int_0^t \sum_{k,l=1}^{n} \sum_{i,j=1}^{m} (B^*(s, Z, \langle e_k \circ e_i\rangle) B(s, Z, \langle e_l \circ e_j\rangle) \times$$

$$\times x, x) \beta^s(e_k, e_i, e_l, e_j) \, ds.$$

By condition I, the integrand converges as $n, m \to \infty$ to $C^\beta(s, Z, Z)$ (the series (5.8) is uniformly convergent). Therefore, applying Theorem 5 of § 9, we see that the series on the right of (5.11) is strongly convergent to a continuous operator-valued martingale with characteristic

$$\int_0^t C^\beta(s, Z, Z) \, ds.$$

For an interval $\Delta = [s, t)$ and $Z \in \mathbf{L}(\mathbf{X})$, we put

$$I_\Delta(Z) = \int_s^t B(u, Z, dY_u(\omega)).$$

$I_\Delta(Z)$ is an $\mathfrak{f}_t$-measurable random operator with values in

$\mathbf{L}_s(\Omega, \mathbf{X})$ (for any Z), with

$$\mathbf{M}(|I_\Delta(Z)x|^2/\mathfrak{f}_s) = \int_s^t C^\beta(u, Z, Z)\,du.$$

We now extend $I_\Delta(Z)$ to random Z such that $Z(\omega)$ is an $\mathfrak{f}_s$-measurable function with values in $\mathbf{L}_s(\Omega, \mathbf{X})$. Obviously, $I_\Delta(Z)$ is independent of the σ-algebra $\mathfrak{f}_s$, but a linear function of Z. If $Z(\omega)$ is $\mathfrak{f}_s$-measurable, then for all n the expression

$$I_\Delta(Z(\omega)P_n) = \sum_{k=1}^n I_\Delta(\langle e_k \circ Z(\omega)e_k\rangle) =$$

$$= \sum_{k=1}^n \sum_{i=1}^\infty (Z(\omega)e_k, e_i)\, I_\Delta(\langle e_k \circ e_i\rangle)$$

is meaningful, since for all k

$$\mathbf{M}\left(\left|\sum_{i=1}^\infty (Z(\omega)e_k, e_i)\, I_\Delta(\langle e_k \circ e_i\rangle)\, x\right|^2 / \mathfrak{f}_s\right) =$$

$$= \sum_{i,j=1}^\infty (Z(\omega)e_k, e_i)(Z(\omega)e_k, e_j) \int_s^t (C^\beta(u, \langle e_k \circ e_i\rangle, \langle e_k \circ e_j\rangle) \times$$

$$\times x, x)\,du = \int_s^t (C^\beta(u, \langle e_k \circ Z(\omega)) e_k\rangle, \langle e_k \circ Z(\omega)e_k\rangle)\, x, x)\,du.$$

Using the last relationship, one readily sees that

$$\mathbf{M}(|I_\Delta(Z(\omega)P_n)x|^2/\mathfrak{f}_s) = \int_s^t (C^\beta(u, Z(\omega)P_n, Z(\omega)P_n)\, x, x)\,ds,$$

$$\mathbf{M}(|I_\Delta(Z(\omega)\,(P_n - P_m))\,x|^2/\mathfrak{f}_s) =$$

$$= \int_0^t (C^\beta(u, Z(\omega)(P_n - P_m), Z(\omega)(P_n - P_m))\, x, x)\,du.$$

Consequently, if $Z(\omega)$ is such that the integral

$$\int_s^t C^\beta(u, Z(\omega), Z(\omega))\,du$$

exists, then one can also define, as a strong limit,

$$I_\Delta(Z(\omega)) = \lim_{n\to\infty} I(Z(\omega)P_n).$$

If, moreover

$$\int_s^t \mathbf{M}\,(C^\beta(u, Z(\omega), Z(\omega))\,x, x)\,du < \infty$$

for all $x \in \mathbf{X}$, then

$$\mathbf{M}\,|I_\Delta(Z(\omega))x|^2 < \infty$$

and for $\tau \in \Delta$ the integral $I_{[s,\tau]}(Z(\omega))$, as a function of τ, is a continuous operator-valued martingale with characteristic

$$\int_s^\tau C^\beta(u, Z(\omega), Z(\omega))\,du.$$

Let $Z_s(\omega)$ be a step function in T_β^2, constant on intervals $[s_k, s_{k+1})$, $k = 0, \ldots,$ where $0 = s_0 < s_1 < \cdots,\ s_n \to \infty$. Then for all k

$$\int_{s_k}^{s_{k+1}} \mathbf{M}C^\beta(s, Z_s(\omega), Z_s(\omega))\,ds = \int_{s_k}^{s_{k+1}} \mathbf{M}C^\beta(s, Z_{s_k}(\omega), Z_{s_k}(\omega))\,ds$$

and so, for $\tau \in [s_k, s_{k+1})$,

$$I_{[s_k,\tau)}(Z_{s_k}(\omega))$$

is defined. Set

$$\int_0^t B(s, Z_s(\omega), dY_s(\omega)) = \sum I_{[s_k\wedge t, s_{k+1}\wedge t)}(Z_{s_k}(\omega))$$

(the sum on the right of this equality involves only finitely many terms). Since $I_{[s_k,t)}(Z_{s_k}(\omega))$, $t \in [t_k, t_{k+1})$ is a continuous operator-valued martingale with characteristic

$$\int_{s_k}^t C^\beta(u, Z_{s_k}(\omega), Z_{s_k}(\omega))\,du = \int_{s_k}^t C^\beta(u, Z_u(\omega), Z_u(\omega))\,du,$$

it follows that the integral thus defined satisfies condition A; that it also satisfies condition B is obvious.

If $Z_t^{(n)}(\omega)$ is a sequence of step functions in T_β^2, such that for all $t > 0, x \in \mathbf{X}$,

$$\lim_{n,m\to\infty} \int_0^t (\mathbf{M}C^\beta(u, Z_u^{(n)}(\omega) - Z_u^{(m)}(\omega), Z_u^{(n)}(\omega) - Z_u^{(m)}(\omega))x, x)\,du = 0, \tag{5.12}$$

then the sequence of martingales

$$\int_0^t B(s, Z_s^{(n)}(\omega), dY_s(\omega))$$

is strongly convergent in mean square (by Theorem 5 of § 9). Let us assume that $Z_t(\omega)$ is a function in T^2_β, such that for all $x \in \mathbf{X}, t > 0$,

$$\lim_{n\to\infty} \int_0^t (\mathbf{M}C^\beta(u, Z_u^{(n)}(\omega) - Z_u(\omega), Z_u^{(n)}(\omega) - Z_u(\omega))\, x, x)\, du = 0. \tag{5.13}$$

Then the sequence $Z_t^{(n)}(\omega)$ satisfies (5.12) and we can put

$$\int_0^t B(s, Z_s(\omega), dY_s(\omega)) = \lim_{n\to\infty} \int_0^t B(s, Z_s^{(n)}(\omega), dY_s(\omega))$$

(the existence of the strong limit on the right follows from (5.12)). It is easily seen that this limit is independent of the choice of a sequence satisfying (5.13).

Let $Z_t(\omega) \in T^2_\beta$. Then the following integral is defined:

$$\int_0^t B(s, (Z_s(\omega)\, e_k, e_l)\, \langle e_k \circ e_l\rangle, dY_s(\omega)) =$$

$$= \int_0^t (Z_s(\omega)\, e_k, e_l)\, B(s, \langle e_k \circ e_l\rangle, dY_s(\omega)).$$

To verify this, it will suffice to construct a sequence of real-valued step functions $\psi_n(s, \omega)$ such that $\psi_n(s, \omega)\, \langle e_k \circ e_l\rangle \in T^2_\beta$ and for all $t > 0$, $x \in \mathbf{X}$,

$$\lim_{n\to\infty} \left(\left\{\mathbf{M} \int_0^t [\psi_n(s, \omega) - (Z_s(\omega)\, e_k, e_l)]^2 \times \right.\right.$$

$$\left.\left. \times\, C^\beta(s, \langle e_k \circ e_l\rangle, \langle e_k \circ e_l\rangle)\, ds\right\} x, x\right) = 0.$$

It follows from the continuity of $(C^\beta(s, \langle e_k \circ e_l\rangle, \langle e_k \circ e_l\rangle)\, x, x)$ that a sufficient condition for this to be true is that

$$\lim_{n\to\infty} \int_0^t \mathbf{M}\, [\psi_n(s, \omega) - (Z_s(\omega)\, e_k, e_l)]^2\, ds = 0. \tag{5.14}$$

The existence of such an $\mathfrak{F}_s$-compatible sequence of step functions $\psi_n(s, \omega)$ is proved by an argument analogous to that used for Lemma 1 in [6, p. 487].

We now define (for the same function $Z_t(\omega)$) the integral

$$\int_0^t B(s, \langle e_k \circ Z_s(\omega) e_k\rangle, dY_s(\omega)).$$

Let $\psi_\alpha(s) = 1$ if $\sup_n \| MC^\beta(s, \langle e_k \circ P_n Z_s(\omega) e_k\rangle, \langle e_k \circ P_n \times \times Z_s(\omega) e_k\rangle)\| \leqslant \alpha$, $\psi_\alpha(s) = 0$ if $\sup_n \| MC^\beta(s, \langle e_k \circ P_n Z_s(\omega) \times \times e_k\rangle, \langle e_k \circ P_n Z_s(\omega) e_k\rangle)\| > \alpha$. For all x,

$$(MC^\beta(s, \langle e_k \circ P_n Z_s(\omega) e_k\rangle, \langle e_k \circ P_n Z_s(\omega) e_k\rangle) x, x) \to$$
$$\to (MC^\beta(s, \langle e_k \circ Z_s(\omega) e_k\rangle, \langle e_k \circ Z_s(\omega) e_k\rangle) x, x),$$

and so the sequence $\| MC^\beta(s, \langle e_k \circ P_n Z_s(\omega) e_k\rangle, \langle e_k \circ P_n Z_s(\omega) e_k\rangle)\|$ is bounded for any s. Consequently, $\varphi_\alpha(s) \uparrow 1$ as $\alpha \to \infty$. Now put

$$\int_0^t B(s, \langle e_k \circ \psi_\alpha(s) Z_s(\omega) e_k\rangle, dY_s(\omega)) = \tag{5.15}$$
$$= \sum_{j=1}^{\infty} \int_0^t B(s, \langle e_k \quad \psi_\alpha(s) (Z_s(\omega) e_k, e_j) e_j\rangle, dY_s(\omega)).$$

The series on the right is convergent; this follows from the fact that if $n < m$ the characteristic of the martingale

$$\sum_{j=n+1}^{m} \int_0^t \psi_\alpha(s) (Z_s(\omega) e_k, e_j) B(s, \langle e_k \circ e_j\rangle, dY_s(\omega))$$

is

$$\sum_{i,j=n+1}^{m} \int_0^t \psi_\alpha(s) (Z_s(\omega) e_k, e_i) (Z_s(\omega) e_k, e_j) \times$$
$$\times C^\beta(s, \langle e_k \circ e_i\rangle, \langle e_k \circ e_j\rangle) ds,$$

and its expectation is

$$\int_0^t \sum_{i,j=n+1}^{m} \psi_\alpha(s)\, \mathbf{M}\,(Z_s(\omega)\, e_k,\, e_i)\,(Z_s(\omega)\, e_k,\, e_j) \times$$

$$\times\, C^\beta(s,\, \langle e_k \circ e_i\rangle,\, \langle e_k \circ e_j\rangle)\, ds.$$

But $\left\| \psi_\alpha(s) \sum_{i,j=n+1}^{m} \mathbf{M}\,(Z_s(\omega)\, e_k,\, e_i)\,(Z_s(\omega)\, e_k,\, e_j) \times\right.$

$$\left.\times\, C^\beta(s,\, \langle e_k \circ e_i\rangle,\, \langle e_k \circ e_j\rangle)\right\| = \psi_\alpha(s)\, \|\mathbf{M}C^\beta(s,\, \langle e_k \circ Z_s(\omega)) \times$$

$$\times\, (P_m - P_n)\, e_k\rangle,\, \langle e_k \circ (P_m - P_n)\, Z_s(\omega)\, e_k\rangle \| \leqslant$$

$$\leqslant \psi_\alpha(s)\, (2\, \|\mathbf{M}C^\beta(s,\, \langle e_k \circ P_m Z_s(\omega)\, e_k\rangle,\, \langle e_k \circ P_m Z_s(\omega)\,\rangle)\| +$$

$$+\, 2\, \|\mathbf{M}C^\beta(s,\, \langle e_k \circ P_n Z_s(\omega)\, e_k\rangle,\, \langle e_k \circ P_n Z_s(\omega)\, e_k\rangle)\|) \leqslant 4\alpha.$$

In addition, for all s and $x \in \mathbf{X}$,

$$\lim_{n,m\to\infty} \sum_{i,j=n+1}^{m} (\mathbf{M}\,(Z_s(\omega)\, e_k,\, e)\,(Z_s(\omega)\, e_k,\, e_j)\, C^\beta(s,\, \langle e_k \circ e_i\rangle,\, \langle e_k \circ$$

$$\circ\, e_j\rangle)\, x,\, x) = 0$$

since the series (5.10) is convergent. Hence, for all $x \in \mathbf{X}, t > 0$,

$$\lim_{n,m\to\infty} \int_0^t \sum_{i,j=n+1}^{m} \psi_\alpha(s)\, \mathbf{M}\,(Z_s(\omega)\, e_k,\, e_i)\,(Z_s(\omega)\, e_k,\, e_j) \times$$

$$\times\, C^\beta(s,\, \langle e_k \circ e_i\rangle,\, \langle e_k \circ e_j\rangle)\, ds = 0.$$

This proves the convergence of the series on the right of (5.15). Now, for $\alpha < \alpha_1$ the characteristic of the martingale

$$\int_0^t B(s,\, \langle e_k \circ [\psi_{\alpha_1}(s) - \psi_\alpha(s)]\, Z_s(\omega)\, e_k\rangle,\, dY_s(\omega))$$

is

$$\int_0^t [\psi_{\alpha_1}(s) - \psi_\alpha(s)]\, C^\beta(s, \langle e_k \circ Z_s(\omega) e_k\rangle, \langle e_k \circ Z_s(\omega) e_k\rangle)\, ds,$$

and this expression converges strongly to zero as $\alpha, \alpha_1 \to \infty$, This implies the existence of the strong limit

$$\lim_{\alpha\to\infty} \int_0^t B(s, \langle e_k \circ \psi_\alpha(s) Z_s(\omega) e_k\rangle, dY_s(\omega)).$$

This limit is by definition

$$\int_0^t B(s, \langle e_k \circ Z_s(\omega) e_k\rangle, dY_s(\omega)); \tag{5.16}$$

it is a martingale with characteristic

$$\int_0^t C^\beta(s, \langle e_k \circ Z_s(\omega) e_k\rangle, \langle e_k \circ Z_s(\omega) e_k\rangle)\, ds,$$

and moreover the integral (5.16) is an additive function of Z.

We now put

$$\int_0^t B(s, Z_s(\omega) P_n, dY_s(\omega)) = \sum_{k=1}^n \int_0^t B(s, \langle e_k \circ Z_s(\omega) e_k\rangle, dY_s(\omega)). \tag{5.17}$$

This integral is a martingale with characteristic

$$\int_0^t \mathbf{M}C^\beta(s, Z_s(\omega) P_n, Z_s(\omega) P_n)\, ds.$$

Putting

$$\bar{\psi}_\alpha(s) = \begin{cases} 1, & \text{if } \sup_n \|C^\beta(s, Z_s(\omega) P_n, Z_s(\omega) P_n)\| \leqslant \alpha, \\ 0, & \text{if } \sup_n \|C^\beta(s, Z_s(\omega) P_n, Z_s(\omega) P_n)\| > \alpha, \end{cases}$$

we define, just as before,

$$\int_0^t B(s, \bar{\psi}_\alpha(s) Z_s(\omega), dY_s(\omega)) = \lim_{n\to\infty} \int_0^t B(s, \bar{\psi}_\alpha(s) \times$$

$$\times Z_s(\omega) P_n, dY_s(\omega)),$$

and then

$$\int_0^t B(s, Z_s(\omega), dY_s(\omega)) = \lim_{\alpha\to\infty} \int_0^t B(s, \overline{\psi}_\alpha(s) Z_s(\omega), dY_s(\omega)),$$

where both limits exist in the sense of strong convergence in mean square. It is easy to see that the integral thus defined satisfies conditions A and B.

Remark. Let $Y^s(\omega)$, be a Gaussian random operator, independent of $Z_s(\omega)$, such that

$$\mathbf{M}(Y^s(\omega)x, y)(Y^s(\omega)u, v) = \beta^s(x, y, u, v).$$

Then by definition of C^β

$$C^\beta(s, Z_s(\omega), Z_s(\omega)) = \mathbf{M}(B^*(s, Z_s(\omega), Y^s(\omega)) \times$$
$$\times B(s, Z_s(\omega), Y^s(\omega))/Z_s(\omega))$$

and so

$$\mathbf{M}C^\beta(s, Z_s(\omega), Z_s(\omega)) = \mathbf{M}B^*(s, Z_s(\omega), Y^s(\omega)) \times$$
$$\times B(s, Z_s(\omega), Y^s(\omega)).$$

Therefore, by condition A,

$$\mathbf{M}\left|\int_0^t B(s, Z_s(\omega)), dY_s(\omega))x\right|^2 = \int_0^t \mathbf{M}\,|B(s, Z_s(\omega), Y^s(\omega))x|^2\,ds$$

or

$$\mathbf{M}\left[\int_0^t B(s, Z_s(\omega), dY_s(\omega))\right]^*\left[\int_0^t B(s, Z_s(\omega), dY_s(\omega))\right] =$$
$$= \int_0^t \mathbf{M}B^*(s, Z_s(\omega), Y^s(\omega))\,B(s, Z_s(\omega), Y^s(\omega))\,ds. \qquad (5.18)$$

Similarly one proves the equality (5.19)

$$\mathbf{M}\left(\int_0^t B(s, Z_s(\omega), dY_s(\omega))x, y\right)\left(\int_0^t B(s, Z_s(\omega), dY_s(\omega))u, v\right) =$$
$$= \int_0^t \mathbf{M}(B(s, Z_s(\omega), Y^s(\omega))x, y)(B(s, Z_s(\omega), Y^s(\omega))u, v)\,ds$$

for all $x, y, u, v \in \mathbf{X}$.

14. Linear Differential Operator Equations

14.1. Definition of a Linear Equation

In this section we shall investigate existence and uniqueness conditions for solutions of the linear stochastic differential equation introduced in § 13 (equation (5.6)), and study the properties of the solutions. The integral form of this equation is

$$X_t(\omega) = X_0(\omega) + \int_0^t A(s, X_s(\omega))\, ds + \int_0^t B(s, X_s(\omega), dY_s(\omega)), \tag{5.20}$$

where $X_0(\omega)$ is a given initial value of the unknown solution $X_t(\omega) \in \mathbf{L}_s(\Omega, \mathbf{X})$, $A(s, Z)$ is a linear function of Z from $[0, \infty) \times \mathbf{L}(\mathbf{X})$ to $\mathbf{L}(\mathbf{X})$, and $B(s, Z_1, Z_2)$ is a linear function of Z_1 and Z_2 from $[0, \infty) \times \mathbf{L}(\mathbf{X}) \times \mathbf{L}(\mathbf{X})$ to $\mathbf{L}(\mathbf{X})$.

Fix some filtration of σ-algebras $\mathfrak{f}_t$ and a Gaussian process with independent increments $Y_t(\omega)$, taking values in $\mathbf{L}_s(\Omega, \mathbf{X})$ and adapted to the filtration $\mathfrak{f}_t$. For the stochastic integral on the right of (5.20) to be meaningful (the integral was defined in § 13, subsection 2), we assume the following conditions to hold:

I. $\mathbf{M}(Y_t(\omega)x, y) = 0, \quad x, y \in X;$

$$\mathbf{M}(Y_t(\omega)x, y)(Y_t(\omega)u, v) = \int_0^t \beta^s(x, y, u, v)\, ds,$$

$(x, y, u, v \in \mathbf{X})$,

where $\beta^s(x, y, u, v)$, $s \geqslant 0$, is a bounded quadrilinear form on $\mathbf{X}^4$, continuous with respect to s.

II. The function

$$C^\beta(s, Z_1, Z_2) = \sum_{k,i,l,j} \beta^s(e_k, e_i, e_l, e_j) B^*(s, Z_1, \langle e_l \circ e_j \rangle) \times$$
$$\times B(s, Z_2, \langle e_k \circ e_i \rangle)$$

is defined, where the series on the right is strongly convergent locally uniformly in s, $\{e_k\}$ is a fixed basis; with this definition the function C^β $(s, Z_1, Z_2$ is (strongly) continuous in s, and also continuous in Z_1 and Z_2 in the norm of $\mathbf{L}(\mathbf{X})$; moreover, for all $x, y \in \mathbf{X}$,

$$\lim_{n,m\to\infty} (C^{\beta}(s, Z_1P_n, Z_2P_m)x, y) = (C^{\beta}(s, Z_1, Z_2)x, y),$$

where $P_n = \sum_{k=1}^{n} \langle e_k \circ e_k \rangle$.

When these conditions are satisfied, the stochastic integral in (5.20) is defined if, for all $t > 0$, $x \in \mathbf{X}$,

$$\int_0^t (\mathbf{M}C^{\beta}(s, X\ (\omega)\, X_s(\omega))\, x, x)\, ds < \infty,$$

where

$$(\mathbf{M}C^{\beta}(s, X_s(\omega),\ X_s(\omega))\, x, x) = \lim_{n,m\to\infty} (\mathbf{M}C^{\beta}(s, X_s(\omega) \times$$

$$\times P_n, X_s(\omega)\, P_m)\, x, x).$$

It will be necessary to substitute a random operator - an element of $\mathbf{L}_s(\Omega, \mathbf{X})$ - into $A(s, Z)$, which is a linear function from $\mathbf{L}(\mathbf{X})$ to $\mathbf{L}(\mathbf{X})$; the following condition will ensure that this substitution is legitimate:

III. $A(s, Z)$ is strongly continuous as a function of s, continuous in the $\mathbf{L}(\mathbf{X})$ -norm as a function of Z, and for all x

$$A(s, Z)\, x = \lim_{n\to\infty} A(s, ZP_n)\, x.$$

We now establish sufficient conditions for the existence of the first integral on the right of (5.20).

Lemma 1. Let $\mathbf{M}\,|X_s(\omega)\, x|^2 < \infty$ for all $x \in \mathbf{X}, s \geqslant 0$,

$$\beta_X(s, t;\ x, y, u, v) = \mathbf{M}\,(X_s(\omega)\, x, y)\,(X_t(\omega)\, u, v)$$

and suppose that the following series converges for all $s, t \geqslant 0$, $x \in \mathbf{X}$:

$$\sum_{j,k,l} (A(s, \langle e_i \circ e_j \rangle)\, x, A(t, \langle e_k \circ e_l \rangle)\, x)\, \beta_X(s, t;\ e_i, e_j, e_k, e_l) =$$

$$= (R(s, t)\, x, x).$$

If the integral

$$\int_0^t \int_0^t (R(\sigma, \tau) x, x)\, d\sigma d\tau$$

exists for all $t > 0$, $x \in \mathbf{X}$, then for all $t > 0$, $x \in \mathbf{X}$, the integral $\int_0^t [A(\tau, X_\tau \times (\omega)) x]\, d\tau$ is defined and is an element of $\mathbf{X}(\Omega)$, i.e.,

$$\int_0^t A(\tau, X_\tau(\omega))\, d\tau \in \mathbf{L}_s(\Omega, \mathbf{X}).$$

Proof. It follows from the lemma of § 13 that, under the assumptions of this lemma,

$$Z_s(\omega) = A(s, X_s, (\omega)) \in \mathbf{L}(\Omega, \mathbf{X}),$$

and moreover

$$\mathbf{M}(Z_s(\omega) x, Z_t(\omega) x) = (R(s, t) x, x).$$

The random process $z(t, \omega) = Z_t(\omega) x$ is measurable,

$$\mathbf{M}(Z(t, \omega), Z(s, \omega)) = \gamma(t, s) = (R(s, t) x, x)$$

and the function $\gamma(t, s)$ is integrable on $[0, T] \times [0, T]$, for any T. Let $\Lambda_n = \{t: \gamma(t, t) < n\}$. Then for all t the integral

$$\int_0 \chi_{\Lambda_n}(s) z(s, \omega)\, ds$$

is defined, since

$$\mathbf{M} \int_0^t \chi_{\Lambda_n}(s) |z(s, \omega)|^2 ds = \int_0^t \chi_{\Lambda_n}(s) \gamma(s, s)\, ds < \infty.$$

It is easily seen that

$$\mathbf{M}\left(\int_0^t \chi_{\Lambda_n}(s) z(s, \omega)\, ds, \int_0^t \chi_{\Lambda_n}(s) z(s, \omega)\, ds \right) =$$

$$= \int_0^t \int_0^t \chi_{\Lambda_n}(\sigma)\, \chi_{\Lambda_m}(\tau)\, \gamma(\sigma, \tau)\, d\sigma d\tau.$$

Since $\chi_{\Lambda_n}(s) \uparrow 1$ as $n \to \infty$, we have

$$\lim_{n,m\to\infty} \mathbf{M}\left(\int_0^t \chi_{\Lambda_n}(s)\, z(s, \omega)\, ds,\ \int_0^t \chi_{\Lambda_m}(s)\, z(s, \omega)\, ds\right) =$$

$$= \int_0^t \int_0^t \gamma(\sigma, \tau)\, d\sigma d\tau$$

and so

$$\lim_{n,m\to\infty} \mathbf{M}\left|\int_0^t \chi_{\Lambda_n}(s)\, z(s, \omega)\, ds - \int_0^t \chi_{\Lambda_m}(s)\, z(s, \omega)\, ds\right|^2 = 0.$$

This implies the existence of the integral

$$\int_0^t z(s, \omega)\, ds = \lim_{n\to\infty} \int_0^t \chi_{\Lambda_n}(s)\, z(s, \omega)\, ds$$

(in the sense of strong convergence in mean square).

14.2. Existence of Uniqueness of Solution

In the theory of stochastic differential equations, it is usually assumed in existence and uniqueness proofs for the solution(s) that the coefficients of the equation satisfy Lipschitz conditions. In the treatment of operator equations with linear coefficients, certain analogs of Lipschitz conditions are employed. Let $C(Z_1, Z_2)$ be a linear function of both its arguments, from $\mathbf{L}(\mathbf{X}) \times \mathbf{L}(\mathbf{X})$ to $\mathbf{L}(\mathbf{X})$, $C(Z_1, Z_2) = C^*(Z_2, Z_1)$ and $C(Z, Z)$ is nonnegative. We shall say that $C(Z_1, Z_2)$ satisfies condition L_γ if $C(Z(\omega), Z(\omega))$ is defined for all $Z(\omega) \in \mathbf{L}_s(\Omega, \mathbf{X})$ such that $\mathbf{M}Z^*(\omega)\, Z(\omega)$ exists, and

$$\|\mathbf{M}C(Z(\omega), Z(\omega))\| \leqslant \gamma \|\mathbf{M}Z^*(\omega)\, Z(\omega)\|.$$

The following is an example of a bilinear function satisfying condition L_γ. Let $S, V \in \mathbf{L}(\mathbf{X})$,

$$C(Z_1, Z_2) = S^* Z_2^* V^* V Z_1 S.$$

Then

$$(\mathbf{M}C(Z(\omega), Z(\omega))x, x) = \mathbf{M}|VZ(\omega)Sx|^2 \leqslant \|V\|^2 \mathbf{M}|Z(\omega)Sx|^2 = \\ = \|V\|^2 (\mathbf{M}Z^*(\omega)Z(\omega)Sx, Sx) \leqslant \|V\|^2 \|\mathbf{M}Z^*(\omega)Z(\omega)\| \|S\|^2 |x|^2.$$

Thus,

$$\|\mathbf{M}C(Z(\omega), Z(\omega))\| = \sup_{|x|=1} (\mathbf{M}C(Z(\omega), Z(\omega))x, x) \leqslant \\ \leqslant \|V\|^2 \|S\|^2 \|\mathbf{M}Z^*(\omega)Z(\omega)\|$$

and $C(Z_1, Z_2)$ satisfies condition L_γ with $\gamma = \|V\|^2\|S\|^2$.

As the sum of bilinear functions satisfying condition L_γ also satisfies a similar condition (with different γ), functions of the form

$$C(Z_1, Z_2) = \sum_k S_k^* Z_2 V_k^* V_k Z_1 S_k$$

will satisfy condition $L_\gamma \left(\gamma = \sum_k \|S_k\|^2 \|V_k\|^2\right)$.

Theorem. Assume that for all $T > 0$ one can exhibit a number $\gamma > 0$ such that the functions $A^*(s, Z_2)A(s, Z_1)$ and $C^\beta(s, Z_1, Z_2)$ satisfy condition L_γ, let conditions I-III hold. If $\mathbf{M}X_0^*(\omega)X_0(\omega)$ is defined and $X_0(\omega)$ is $\mathfrak{f}_0$-measurable, then equation (5.20) has an $\mathfrak{f}_t$-compatible solution $X_t(\omega)$ such that $\mathbf{M}X_t^*(\omega)X_t(\omega)$ exists and $\|\mathbf{M}X_t^*(\omega)X_t(\omega)\|$ is locally bounded. There is exactly one solution with these properties, up to stochastic equivalence: if $\overline{X}_t(\omega)$ is any other, then $\forall x \in \mathbf{X}$, $t > 0$,

$$\{\overline{X}_t(\omega)x = X_t(\omega)x\} = 1.$$

Proof. We first prove uniqueness. If $X_t(\omega)$ and $\overline{X}_t(\omega)$ are solutions of (5.20), then

$$X_t(\omega) - \overline{X}_t(\omega) = \int_0^t A(s, X_s(\omega) - \overline{X}_s(\omega))\,ds + \tag{5.21}$$

$$+ \int_0^t B(s, X_s(\omega) - \overline{X}_s(\omega), dY_s(\omega)).$$

Denoting $X_t(\omega) - \overline{X}_t(\omega) = Z_t(\omega)$, we obtain $(t \leqslant T)$

$$\mathbf{M}|Z_t(\omega)x|^2 \leqslant 2\mathbf{M}\left|\int_0^t A(s, Z_s(\omega))x\,ds\right|^2 +$$

$$+ 2\mathbf{M}\left|\int_0^t B(s, Z_s(\omega), dY_s(\omega))x\right|^2 \leqslant 2t\int_0^t \mathbf{M}(A^*(s, Z_s(\omega)) \times$$

$$\times A(s, Z_s(\omega))x, x)\,ds + 2t\int_0^t (\mathbf{M}C^\beta(s, Z_s(\omega), Z_s(\omega))x, x)\,ds \leqslant$$

$$\leqslant 2t\int_0^t \gamma\|\mathbf{M}Z_s^*(\omega)Z_s(\omega)\||x|^2\,ds + 2\gamma\int_0^t \|\mathbf{M}Z_s^*(\omega)Z_s(\omega)\| \times$$

$$\times |x|^2\,ds \leqslant (2T+2)\gamma\int_0^t \|\mathbf{M}Z_s^*(\omega)Z_s(\omega)\|\,ds\,|x|^2.$$

Thus, for $|x| = 1$, $t \leqslant T$,

$$(\mathbf{M}Z_t^*(\omega)Z_t(\omega)x, x) \leqslant (2T+2)\gamma\int_0^t \|\mathbf{M}Z_s^*(\omega)Z_s(\omega)\|\,ds.$$

Evaluating the supremum of the expression on the left over $X\,|x| = 1$, we see that for some $\lambda < \infty$

$$\|\mathbf{M}Z_t^*(\omega)Z_t(\omega)\| \leqslant \lambda\int_0^t \|\mathbf{M}Z_s^*(\omega)Z_s(\omega)\|\,ds. \tag{5.22}$$

It follows from this inequality that $\|\mathbf{M}Z_t^*(\omega)Z_t(\omega)\| = 0$. This proves uniqueness. As usual in the theory of stochastic differential equations, the existence proof utilizes successive approximations. We put

$$X_t^0(\omega) = X_0(\omega), \tag{5.23}$$

$$X_t^{(n}(\omega) = X_0(\omega) + \int_0^t A(s, X_s^{(n-1)}(\omega))\,ds +$$

$$+ \int_0^t B(s, X_s^{(n-1)}(\omega), dY_s(\omega)), \quad n \geqslant 1;$$

$$Z_t^{(n)}(\omega) = X_t^{(n)}(\omega) - X_t^{(n-1)}(\omega), \quad n > 0, \quad Z_t^{(0)}(\omega) = X_0(\omega).$$

The existence of the integrals on the right of (5.23) follows from conditions I-III and from the fact that A^*A, C^β satisfy condition L_γ on every finite interval. Subtracting consecutive equations in (5.23), we obtain the recurrence relation

$$Z_t^{(n)}(\omega) = \int_0^t A(s, Z_s^{(n-1)}(\omega))\, ds + \int_0^t B(s, Z_s^{(n-1)}(\omega),\; dY_s(\omega)),$$

whence, following exactly the same procedure by which (5.22) was derived from (5.21), we deduce that for some λ, depending on T,

$$\| \mathbf{M}\, [Z_t^{(n)}(\omega)]^* Z_t^{(n)}(\omega) \| \leqslant \lambda \int_0^t \| \mathbf{M}\, (Z_s^{(n-1)}(\omega))^* Z_s^{(n-1)}(\omega) \|\, ds, \qquad t \leqslant T.$$

Consequently,

$$\| \mathbf{M}\, (Z_t^{(n)}(\omega))^* Z_t^{(n)}(\omega) \| \leqslant \frac{(\lambda t)^n}{n!} \| \mathbf{M} X_0^*(\omega)\, X_0(\omega) \|. \tag{5.24}$$

Since

$$X_t^{(n)}(\omega) = \sum_{k=0}^{n} Z_t^{(k)}(\omega)$$

and the series on the right is strongly convergent in view of (5.24), it follows that the sequence $X_t^{(n)}(\omega)$ has a strong limit

$$X_t(\omega) = \sum_{k=0}^{\infty} Z_t^{(k)}(\omega).$$

It is easy to see that, in the sense of strong operator convergence,

$$\int_0^t A(s, X_s^{(n)}(\omega))\, ds \to \int_0^t A(s, X_s(\omega))\, ds;$$

$$\int_0^t B(s, X_s^{(n)}(\omega),\; dY_s(\omega)) \to \int_0^t B(s, X_s(\omega),\; dY_s(\omega)).$$

Therefore, going to the limit in (5.23), we obtain (5.20). That $\| \mathbf{M} X_t^*(\omega)\, X_t(\omega) \|$ is locally bounded follows from the inequality

$$| \mathbf{M} X_t(\omega)\, x |^2 = \left| \mathbf{M} \sum_{k=0}^{\infty} Z_t^{(k)}(\omega)\, x \right|^2 \leqslant \mathbf{M} \left| \sum_{k=0}^{\infty} 2^{-k} (2^{k} Z_t^{(k)}(\omega)\, x) \right|^2 \leqslant$$

$$\leqslant \sum_{k=0}^{\infty} 4^{-k} \sum_{k=0}^{\infty} 4^k \mathbf{M} \,|\, Z_t^{(k)}(\omega)\, x \,|^2 \leqslant$$

$$\leqslant \frac{4}{3} \sum_{k=0}^{\infty} \frac{(4\lambda t)^k}{k!} \mathbf{M} \,\|\, X_0^*(\omega)\, X_0(\omega) \,\|$$

(here we have used (5.24)). ■

14.3. Linear Transformations of Solutions

Let $X_t(\omega)$ be a solution of an equation (5.20) whose coefficients satisfy the assumptions of the theorem. Let us assume that $R(Z)$ is an invertible linear function from $\mathbf{L}(\mathbf{X})$ to $\mathbf{L}(\mathbf{X})$, satisfying the conditions:

(1) $R(Z(\omega))$ is defined for all $Z(\omega) \in \mathbf{L}_s(\Omega, \mathbf{X})$;

(2) $R^*(Z_2)\, R(Z_1)$ satisfies condition L_γ for some γ.

Consider the function $\hat{X}_t(\omega) = R(X_t(\omega))$. We claim that

$$\hat{X}_t(\omega) = R(X_0(\omega)) + \int_0^t R(A(s, X_s(\omega))\, ds + \qquad (5.25)$$

$$+ \int_0^t R(B(s, X_s(\omega), dY_s(\omega))).$$

To see that this is true, we must verify that ordinary and stochastic integration commute with the operation R. We shall do this for the stochastic integral. We first note that

$$\int_0^t R(B(s, X_s(\omega), dY_s(\omega))\, P_n) = R\left(\int_0^t (B(s, \mathbf{X}_s(\omega), dY_s(\omega))\, P_n \right), \qquad (5.26)$$

since R is continuous in norm, and

$$\int_0^t B(s, X_s(\omega), dY_s(\omega))\, P_n$$

is the limit (in the operator norm) of integral sums and stochastic integrals defined via these sums,

$$\int_0^t B(s, X_s^{(m)}(\omega), dY_s(\omega))\, P_n,$$

and for these integrals $\int_0^t B(s, X_s(\omega), dY_s(\omega))$ is the strong limit of $\int_0^t B(s, X_s^{(m)}(\omega), dY_s(\omega))$ (the construction of the integral is described in § 13, subsection 2). Letting $n \to \infty$ in (5.26), we obtain the required relationship. Now let $R^{-1}(Z)$ denote the inverse function to R. Then, putting

$$\hat{A}(s, Z) = R(A(s, R^{-1}(Z))), \quad \hat{B}(s, Z_1, Z_2) =$$

$$= R(B(s, R^{-1}(Z_1), Z_2)),$$

we can rewrite (5.25) as (5.27)

$$\hat{X}_t(\omega) = \hat{X}_0(\omega) + \int_0^t \hat{A}(s, \hat{X}_s(\omega))\, ds + \int_0^t \hat{B}(s, \hat{X}_s(\omega), dY_s(\omega)).$$

Thus, by using R we have transformed one linear equation (5.20) into another (5.27). The coefficients $\hat{A}$ and $\hat{B}$ satisfy the assumptions of the theorem if $R^{-1}(Z)$ in turn satisfies the two conditions enunciated at the beginning of this subsection.

More important transformations are obtained if R is also dependent on t.

Lemma 2. Let $X_t(\omega)$ be a solution of equation (5.20) and $R_t(Z), t \geqslant 0$, a linear function from $\mathbf{L}(\mathbf{X})$ to $\mathbf{L}(\mathbf{X})$, continuous in norm, such that the following conditions hold:

(a) $R_t(Z)$ is invertible as a function of Z;

(b) the strong derivative $\frac{d}{dt} R_t(Z) = R_t'(Z)$, exists, $R_t'(Z)$ is (strongly) continuous in t, Z;

(c) the functions $R_t(Z)$, $R_t^{-1}(Z)$ and $R_t'(Z)$ are defined for any $Z = Z(\omega) \in \mathbf{L}_s(\Omega, \mathbf{X})$ such that $\mathbf{M}Z^*(\omega) Z(\omega)$; exists;

(d) the function $R_t^*(Z_2) R_t(Z_1)$ satisfies condition L_γ $(\gamma > 0)$;

(e) for any $\varepsilon > 0$ there exists $\delta > 0$ such that when $|s - t| < \delta$ the operator-valued function

$$[R_t - R_s]$$

satisfies condition L_ε.

Then the function $\hat{X}_t(\omega) = R(X_t(\omega))$ satisfies the stochastic equation (5.27), with

$$\hat{A}(t, Z) = R_t'(R_t^{-1}(Z)) + R_t(A(t, R_t^{-1}(Z)));$$

$$\hat{B}(t, Z_1, Z_2) = R_t(B(t, R_t^{-1}(Z_1), Z_2)).$$

Proof. First let

$$R_t = R_0 + tR_1, \quad R_t' = R_1.$$

Then

$$R_t(X_t(\omega)) = R_0(X_t(\omega)) + tR_1(X_t(\omega)).$$

Using formula (5.25) for R_0 and R_1, we obtain

$$R_t(X_t(\omega)) = R_0(X_0(\omega)) + \int_0^t R_0(A(s, X_s(\omega)))\,ds + \tag{5.28}$$

$$+ \int_0^t R_0(B(s, X_s(\omega), dY_s(\omega))) + tR_1(X_0(\omega)) +$$

$$+ t\int_0^t R_1(A(s, X_s(\omega)))\,ds + t\int_0^t R_1(B(s, X_s(\omega), dY_s(\omega))).$$

It is readily shown that

$$t\int_0^t R_1(A(s, X_s(\omega)))\,ds = \int_0^t sR_1(A(s, X_s(\omega)))\,ds + \tag{5.29}$$

$$+ \int_0^t \int_0^\tau R_1(A(s, X_s(\omega)))\,ds\,d\tau;$$

$$t\int_0^t R_1(B(s, X_s(\omega), dY_s(\omega))) = \int_0^t sR_1(B(s, X_s(\omega), dY_s(\omega))) +$$

$$+ \int_0^t \int_0^\tau R_1(B(s, X_s(\omega), dY_s(\omega)))\,d\tau \tag{5.30}$$

(to verify these relationships it suffices to reverse the order of integration in the double integrals on the right). Substituting these relationships into (5.28), we find that

$$R_t(X_t(\omega)) = R_0(X_0(\omega)) + \int_0 [R_0(A(s, X_s(\omega))) +$$

$$+ sR_1(A(s, X_s(\omega))]\, ds + \int_0^t [R_0(B(s, X_s(\omega), dY_s(\omega))) +$$

$$+ sR_1(B(s, X_s(\omega), dY_s(\omega)))] + \int_0^t R_1(X_0(\omega) +$$

$$+ \int_0^s A(\tau, X_\tau(\omega))\, d\tau + \int_0^s B(\tau, X_\tau(\omega), dY_\tau(\omega)))\, ds.$$

In terms of the notation R_t and R_t' this relationship can be rewritten as

$$R_t(X_t(\omega)) = R_0(X_0(\omega)) + \int_0^t [R_s(A(s, X_s(\omega))) +$$

$$+ R_s'(X_s(\omega))]\, ds + \int_0^t R_s(B(s, X_s(\omega), dY_s(\omega))).$$

Let $0 = t_0^{(n)} < t_1^{(n)} < \cdots < t_n^{(n)} = t$,

$$R_t^{(n)}(Z) = R_{t_k^{(n)}}(Z) + \frac{t - t_k^{(n)}}{t_{k+1}^{(n)} - t_k^{(n)}} [R_{t_{k+1}^{(n)}}(Z) - R_{t_k^{(n)}}(Z)]$$

for $t \in [t_k^{(n)}, t_{k+1}^{(n)}]$.

By what we have proved, for any interval $[t_k^{(n)}, t_{k+1}^{(n)}]$:

$$R_{t_{k+1}^{(n)}}^{(n)}(X_{t_{k+1}^{(n)}}(\omega)) - R_{t_k^{(n)}}^{(n)}(X_{t_k^{(n)}}(\omega)) =$$

$$= \int_{t_k^{(n)}}^{t_{k+1}^{(n)}} [R_s^{(n)}(A(s, X_s(\omega))) + R_s^{(n)'}(X_s(\omega))]\, ds +$$

$$+ \int_{t_k^{(n)}}^{t_{k+1}^{(n)}} R_s^{(n)}(B(s, X_s(\omega), dY_s(\omega))).$$

Adding these equalities, we obtain

$$R_t^{(n)}(X_t(\omega)) - R_0^{(n)}(X_0(\omega)) = \int_0^t [R_s^{(n)}(A(s, X_s(\omega))) + \qquad (5.31)$$

$$+ R_s^{(n)'}(X_s(\omega))]\, ds + \int_0^t R_s^{(n)}(B(s, X_s(\omega), dY_s(\omega))).$$

We now let $n \to \infty$ and $\max_k (t_{k+1}^{(n)} - t_k^{(n)}) \to 0$. In view of our assumptions and the equality

$$R_s^{(n)\prime} = \frac{1}{t_{k+1}^{(n)} - t_k^{(n)}} [R_{t_{k+1}^{(n)}} - R_{t_k^{(n)}}],$$

we have

$$[R_s^{(n)}(A(s, X_s(\omega))) + R_s^{(n)\prime}(X_s(\omega))] \to R_s(A(s, \dot{X}_s(\omega))) + \\ + R_s'(X_s(\omega))$$

strongly in mean square and uniformly in s. We may therefore perform the limit passage under the integral sign in the first integral on the right of (5.31). We claim that this is also true for the second integral. Put

$$Q_s^{(n)} = R_s - R_s^{(n)}.$$

By condition (e), the operator-valued function $Q_s^{(n)*} Q_s^{(n)}$ satisfies condition L_ε for sufficiently large n. Therefore, if $\hat{Y}_s(\omega)$ obeys a Gaussian distribution for which

$$\mathbf{M}(\hat{Y}_s(\omega)x, y)(\hat{Y}_s(\omega)u, v) = \beta^s(x, y, u, v)$$

(see condition (1)) and $\hat{Y}_s(\omega)$ is independent of $X_s(\omega)$, then

$$\|\mathbf{M}\,]Q_s^{(n)}(B(s, X_s(\omega), \hat{Y}_s(\omega)))]^* Q_s^{(n)}(B(s, X_s(\omega), \hat{Y}_s(\omega)))\| \leqslant$$
$$\leqslant \varepsilon \|\mathbf{M}B^*(s, X_s(\omega), \hat{Y}_s(\omega)) B(s, X_s(\omega), \hat{Y}_s(\omega))\| =$$
$$= \varepsilon \|\mathbf{M}\mathbf{M}(B^*(s, X_s(\omega), \hat{Y}_s(\omega)) B(s, X_s(\omega), \hat{Y}_s(\omega)) \mid X_s(\omega))\| =$$
$$= \varepsilon \|\mathbf{M}C^\beta(s, X_s(\omega), X_s(\omega))\| \leqslant \varepsilon\gamma \|\mathbf{M}X_s^*(\omega) X_s(\omega)\|.$$

Therefore, by (5.19), we obtain

$$\left\| \mathbf{M} \left[\int_0 Q_s^{(n)}(B(s, X_s(\omega), d\hat{Y}_s(\omega))) \right]^* \times \right.$$
$$\left. \times \left[\int_0^t Q_s^{(n)}(B(s, X_s(\omega), d\hat{Y}_s(\omega))) \right] \right\| \leqslant$$
$$\leqslant \left\| \int_0^t \mathbf{M}\,[Q_s^{(n} (B(s, X_s(\omega), \hat{Y}_s(\omega)))]^* Q_s^{(n)}(B(s, X_s(\omega), \hat{Y}_s(\omega)))\, ds \right\| \leqslant$$
$$\leqslant \varepsilon\gamma \int_0^t \|\mathbf{M}X_s^*(\omega) X_s(\omega)\|\, ds.$$

Consequently,

$$\lim_{n\to\infty}\left\| \mathbf{M}\left[\int_0^t Q_s^{(n)}(B(s, X_s(\omega), dY_s(\omega)))\right]^* \times \right.$$

$$\left. \times \int_0^t Q_s^{(n)}(B(s, X_s(\omega), dY_s(\omega)))\right\| = 0.$$

This proves that it is legitimate to evaluate the limit within the stochastic integral in (5.31). Proceeding to the limit accordingly, we obtain

$$R_t(X_t(\omega)) = R_0(X_0(\omega)) + \int_0^t [R_s'(X_s(\omega)) + \tag{5.32}$$

$$+ R_s(A(s, X_s(\omega)))]\, ds + \int_0^t R_s(B(s, X_s(\omega), dY_s(\omega))),$$

whence follows the proof of the lemma. ■

Choose the function $R_s(Z)$ so that

$$R_s'(Z) + R_s(A(s, Z)) = 0 \tag{5.33}$$

for all $Z \in \mathbf{L}(\mathbf{X})$. Then the integrand of the ordinary integral in (5.32) vanishes, while $R_t(X_t(\omega))$ becomes a martingale. We shall now see how to find a function R_s satisfying (5.33). Let $\mathbf{L}_{\mathrm{I}}(\mathbf{X})$ denote the space of linear functions $R(Z)$ from $\mathbf{L}(\mathbf{X})$ to $\mathbf{L}(\mathbf{X})$ and $\mathbf{L}_{\mathrm{II}}(\mathbf{X})$ the space of linear functions from $\mathbf{L}_{\mathrm{I}}(\mathbf{X})$ to $\mathbf{L}_{\mathrm{I}}(\mathbf{X})$. We norm these space in the neutral way:

if $R \in \mathbf{L}_{\mathrm{I}}(\mathbf{X})$, we put $\|R\|_{\mathrm{I}} = \sup_{\|Z\|\leqslant 1}\|R(Z)\|$,

if $\mathfrak{a} \in \mathbf{L}_{\mathrm{II}}(\mathbf{X})$, we put $\|\mathfrak{a}\|_{\mathrm{II}} = \sup_{\|R\|_{\mathrm{I}}\leqslant 1}\|\mathfrak{a}(R)\|_{\mathrm{I}}$.

Let $A(Z) \in \mathbf{L}_{\mathrm{I}}(\mathbf{X})$. We then define the following linear operation in $\mathbf{L}_{\mathrm{I}}(\mathbf{X})$:

$$[\mathfrak{a}(R)](Z) = R(A(Z)).$$

With this definition,

$$\|\mathfrak{a}\|_{\mathrm{II}} = \sup_{\|R\|_{\mathrm{I}}\leqslant 1}\|R(A(Z))\| = \sup_{\|R\|_{\mathrm{I}}\leqslant 1}\ \sup_{\|Z\|\leqslant 1}\|R(A(Z))\| \leqslant \|A\|.$$

Let $\mathfrak{a}_s(R) = R(A(s, \cdot))$. Equation (5.33) may be regarded as a linear equation in the Banach space $\mathbf{L}_{\mathrm{I}}(\mathbf{X})$:

$$\frac{d}{ds} R_s + \mathfrak{a}_s (R_s) = 0. \tag{5.34}$$

As follows from the general theory of linear equations in Banach spaces, if $\mathfrak{a}_s$ is bounded and continuous in s in the norm $\| \cdot \|_{1}$, then equation (5.34) has a unique solution for any initial condition R_0. Denoting the multiplication in $\mathbf{L}_1 (\mathbf{X})$ by $*$, we can express the operator $\mathfrak{a}_s$ in the form

$$\mathfrak{a}_s (R) = R * A (s, \cdot),$$

i.e., the operator of multiplication by a fixed element of $\mathbf{L}_1 (\mathbf{X})$ (that is to say, by a fixed linear operator). Hence the solutions of equation (5.34) generate a non-homogeneous semi-group: if $R_{t,s}$ is a solution of equation (5.34) $(s \geqslant t)$ satisfying the initial condition $R_{t,t} = I$, where I is the identity operator in $\mathbf{L}_1 (\mathbf{X})$, then for $t < u < s$ we have $R_{t,s} = R_{t,u} * R_{u,s}$. Since the solution is continuous in $\| \cdot \|_1$, it follows that if $s - t$ is sufficiently small, then $R_{t,s}$ is close to the identity and therefore invertible; hence the operator $R_s = R_{0,s}$ satisfying (5.34) is invertible for all $s > 0$.

Transformation of the solution of equation (5.20) to a martingale is particularly simple when $A (s, Z)$ is independent of s. Then equation (5.34) admits the solution

$$R_s = \exp \{- sA (\cdot)\} = I + \sum_{n=1}^{\infty} \frac{(- s)^n}{n!} A^{*n} (\cdot), \tag{5.35}$$

where

$$A^{*n} (Z) = A (A^{*(n-1)} (Z)), \quad A^{*1} (Z) = A (Z).$$

14.4. Equations for Moments of the Solution of a Stochastic Equation

We shall consider only equations for the first two moment functions. Let $S_t = \mathbf{M}X_t (\omega)$ (this function is defined since, under the assumptions of the theorem, $\mathbf{M}X_t^* (\omega) X_t(\omega)$ exists). Evaluating the expectation of both sides of equation (5.20) and noting that the second integral is a martingale, we get

$$S_t = S_0 + \int_0^t \mathbf{M}A(s, X_s(\omega))\,ds =$$

$$= S_0 + \int_0^t A(s, \mathbf{M}X_s(\omega))\,ds = S_0 + \int_0^t A(s, S_s)\,ds.$$

That the order of the operations of evaluation of expectation and integration with respect to s may be interchanged follows from the inequality

$$\mathbf{M}|A(s, X_s(\omega))x| \leqslant \sqrt{\mathbf{M}|A(s, X_s(\omega))x|^2} \leqslant$$
$$\leqslant |x| \sqrt{\gamma \|\mathbf{M}X_s^*(\omega) X_s(\omega)\|}.$$

The equality

$$\mathbf{M}A(X_s(\omega)) = A(\mathbf{M}X_s(\omega))$$

is valid if A^*A satisfies condition L_γ, since

$$\mathbf{M}A(X_s(\omega) P_n)x = A(\mathbf{M}X_s(\omega) P_n)x,$$

and it is legitimate to let $n \to \infty$ on both sides of this equality (on the right - because $\mathbf{M}X_s(\omega) \in \mathbf{L}(\mathbf{X})$; on the left - because $A(X_s(\omega)P_n)x \to A(X_s(\omega))x$ strongly in probability and

$$\mathbf{M}|A(X_s(\omega)P_n)x|^2 \leqslant \gamma |x|^2 \|\mathbf{M}P_n X_s^*(\omega) X_s(\omega) P_n\| \leqslant$$
$$\leqslant \gamma |x|^2 \|\mathbf{M}X_s^*(\omega) X_s(\omega)\|.$$

Thus the operator S_t satisfies a linear differential operator equation

$$\frac{d}{dt} S_t = A(t, S_t), \quad S_t = S_0 \text{ at } t = 0.$$

To evaluate the second moment, we transform the solution of the equation to martingale form. Let R_s be a function in $\mathbf{L}_1(\mathbf{X})$, satisfying the condition $R_0 = I$ and equation (5.33). Then

$$R_t(X_t(\omega)) = X_0(\omega) + \int_0^t R_s(B(s, X_s(\omega), dY_s(\omega))).$$

Putting $\hat{X}_t(\omega) = R_t(X_t(\omega))$, $\hat{B}(s, Z_1, Z_2) = R_s(B(s, R_s^{-1}Z_1, Z_2))$, $\hat{\beta}(s, x, y, u, v) = \mathbf{M}(\hat{X}_s(\omega)x, y)(\hat{X}_s(\omega)u, v)$ and using equality (5.19) (see § 13), we obtain

$$\hat{\beta}(t, x, y, u, v) = \mathbf{M}(X_0(\omega)x, y)(X_0(\omega)u, v) +$$

$$+ \int_0^t \mathbf{M}(\hat{B}(s, \hat{X}_s(\omega), Y^s(\omega))x, y)(\hat{B}(s, \hat{X}_s(\omega), Y^s(\omega))u, v)\, ds.$$

Differentiating this relation with respect to t and using the equality

$$(B(s, \hat{X}_s(\omega), Y^s(\omega))x, y) = \sum_{i,k} (\hat{B}(s, \langle e_i \circ e_k \rangle, Y^s(\omega))x, y) \times$$

$$\times (\hat{X}_s(\omega)e_i, e_k),$$

we get (5.36)

$$\frac{\partial}{\partial t}\hat{\beta}(t, x, y, u, v) = \sum_{i,j,k,l} [\mathbf{M}(\hat{B}(s, \langle e_i \circ e_k \rangle, Y^s(\omega))x, y) \times$$

$$\times (\hat{B}(s, \langle e_j \circ e_l \rangle, Y^s(\omega))u, v)]\hat{\beta}(t, e_i, e_k, e_j, e_l).$$

If x, y, u, v are elements of the basis $\{e_k\}$, equation (5.36) may be interpreted as an infinite linear system of differential equations. Let Φ^{IV} be the space of bounded quadrilinear forms β on $\mathbf{X}^4$ such that

$$\beta(x, y, u, v) = \beta(u, v, x, y), \sup_{|x|\leqslant 1} \sum_k |\beta(x, e_k, x, e_k)| < \infty$$

(the second moment function for a strong operator $Z(\omega)$, with $\mathbf{M}|Z(\omega)x|^2 < \infty$ satisfies these conditions). We denote

$$\gamma^s_{ikjl}(x, y, u, v) = \mathbf{M}(\hat{B}(s, \langle e_i \circ e_k \rangle, Y^s(\omega))x, y) \times$$

$$\times (\hat{B}(s, \langle e_j \circ e_l \rangle, Y^s(\omega))u, v).$$

Then, if the series

$$\sum_{i,j,k,l} \beta(e_i, e_k, e_j, e_l) \gamma^s_{i_k j^l}$$

is convergent for all $\beta \in \Phi^{IV}$ and is a form in Φ^{IV}, we can define a bounded operator on

$$\mathfrak{a}^s \beta = \sum \beta(e_i, e_k, e_j, e_l) \gamma^s_{ikjl}$$

and equation (5.36) may be interpreted as a linear equation

in Φ^{IV}:

$$\frac{d}{ds}\hat{\beta}_s = \mathfrak{a}^s\hat{\beta}_s,\ \beta_0(x, y, u, v) = \mathbf{M}(X_0(\omega)x, y)(X_0(\omega)u, v).$$

$$\beta_X(s, x, y, u, v) = \mathbf{M}(X_s(\omega)x, y)(X_s(\omega)u, v).$$

$$\beta_X(s, x, y, u, v) = \mathbf{M}(R_s^{-1}(X_s(\omega))x, y)(R_s^{-1}(X_s(\omega))u, v) =$$
$$= \sum_{i,k,j,l}(R_s^{-1}(\langle e_i \circ e_k\rangle)x, y)(R_s^{-1}(\langle e_j \circ e_l\rangle)u, v) \times$$
$$\times \hat{\beta}(s, e_i, e_k, e_j, e_l) \tag{5.37}$$

(this series converges if $(R_s^{-1})^* R^{-1}$ satisfies condition L_γ).

15. Continuous Stochastic Semigroups

15.1. Solutions of Simple Linear Equations - Stochastic Semigroups

Consider an equation of the type

$$dX_t(\omega) = A_tX_t(\omega)\,dt + dY_t(\omega)X_t(\omega), \tag{5.38}$$

where A_t is a locally bounded measurable function with values in $\mathbf{L}(\mathbf{X})$ and $Y_t(\omega)$ a Gaussian process with independent increments, such that

$$\mathbf{M}(dY_t(\omega)x, y) = 0,\ \mathbf{M}(dY_t(\omega)x, y)(dY_t(\omega)u, v) =$$
$$= \beta^t(x, y, u, v)\,dt. \tag{5.39}$$

This equation is of the same form as (5.6) (§ 13), with $A(t, Z) = A_t(Z)$, $B(t, Z_1, Z_2) = Z_2^*Z_1$. In our case,

$$\mathbf{M}|A_tZ(\omega)x|^2 \leqslant \|A_t\|^2\mathbf{M}|Z(\omega)x|^2,$$

$$C^\beta(t, Z_1, Z_2) = Z_2^*\mathbf{M}Y^{t*}(\omega)Y^t(\omega)Z_1 = Z_2^*R_tZ_1,$$

where $(R_tx, x) = \mathbf{M}|Y^t(\omega)x|^2$, (in the notation of § 13).
If R_t is a bounded function in $\mathbf{L}(\mathbf{X})$, then

$$\mathbf{M}(C^\beta(t, Z(\omega), Z(\omega))x, x) \leqslant \|R_t\|\mathbf{M}|Z(\omega)x|^2.$$

Thus, both A^*A, and C^β satisfy condition L_γ. By the theorem of § 14, equation (5.38) has a unique solution on any interval $[s, \infty)$, $(s > 0)$ for any initial condition $X_s(\omega)$ which is independent of the process $\{Y_t(\omega) - Y_s(\omega),\ t \geqslant s\}$. Let $X_t^s(\omega)$ denote the solution of equation (5.38) on $[s, \infty)$ such that

$X_s^s(\omega) = I$. The most important properties of the operator-valued function $\{X_t^s(\omega),\ 0 \leqslant s \leqslant t\}$ are as follows:

1. Let $\mathfrak{f}_t^s$ be the σ-algebra generated by the functions $Y_u(\omega) - Y_s(\omega),\ s \leqslant u \leqslant t$. Then $X_t^s(\omega)$ is $\mathfrak{f}_t^s$-measurable and hence, for any $0 = t_0 < t_1 < \cdots < t_{n+1}$, the functions

$\{X_{t_{k+1}}^{t_k}(\omega),\ k = 0, \ldots, n\}$ are mutually independent (since the σ-algebras $\mathfrak{f}_{t_{k+1}}^{t_k}$ are independent).

2. The following equality is true with probability 1 $(s < u < t)$:

$$X_t^u(\omega)\, X_u^s(\omega) = X_t^s(\omega). \tag{5.40}$$

Indeed, suppose that for $t > u$ (u and s being fixed) $Z_t(\omega) = X_t^u(\omega)\, X_u^s(\omega)$. Then

$$Z_t(\omega) = X_u^s(\omega) + \left(\int_u^t A_\tau X_\tau^u(\omega)\, d\tau\right) X_u^s(\omega) +$$
$$+ \left(\int_u^t dY_\tau(\omega)\, X_\tau^u(\omega)\right) X_u^s(\omega).$$

It is easy to see that the order of the operations of stochastic or ordinary integration and application of an $\mathfrak{f}_u^0$-measurable operator may be intercganged, as the appropriate interchange is certainly legitimate in the integral sums that approximate either type of integral. Therefore,

$$Z_t(\omega) = X_u^s(\omega) + \int_u^t A_\tau Z_\tau(\omega)\, d\tau + \int_u^t dY_\tau Z_\tau(\omega)\, d\tau. \tag{5.41}$$

Now, for $t \geqslant 0$

$$X_t^s(\omega) = I + \int_s^t A_\tau X_\tau^s(\omega)\, d\tau + \int_s^t dY_\tau(\omega)\, X_\tau^s(\omega) = \tag{5.42}$$
$$= X_u^s(\omega) + \int_u^t A_\tau X_\tau^s(\omega)\, d\tau + \int_u^t dY_\tau(\omega)\, X_\tau^s(\omega).$$

Comparing (5.41), (5.42) and recalling that the solution is unique, we obtain $Z_t(\omega) = X_t^s(\omega),\ t \geqslant u$, thus proving (5.40).

3. The expectation of any solution of the type we are considering exists: $\mathbf{M}\mathbf{X}_t^s(\omega) = S_t^s \in \mathbf{L}(\mathbf{X})$; with this notation,

$S_t^s = S_t^u S_u^s$ for $s \leqslant u \leqslant t$, S_t^u satisfies the integral equation

$$S_t^u = I + \int_u^t A_\tau S_\tau^u d\tau, \tag{5.43}$$

and therefore S_t^u is continuous in the $\mathbf{L}(\mathbf{X})$ -norm as a function of t and has a continuous inverse (as a function of t). Since $S_t^u = S_t^0 [S_u^0]^{-1}$, it follows that S_t^u is continuous (in the norm $\| \cdot \|$) in both variables. All these properties of S_t^u follow from equation (5.43), which may be derived from (5.42) by evaluating and setting $s = u$.

4. Let

$$\mathbf{M} | dY_t(\omega) x |^2 = (R_t x, x) dt,$$

where R_t is a locally bounded function in $\mathbf{L}(\mathbf{X})$. Then, putting $\gamma(\tau) = 2 \| A_\tau \| + \| R_\tau \|$, we obtain

$$\| \mathbf{M} (X_t^s(\omega))^* X_t^s(\omega) \| \leqslant \exp \left\{ \int_s \gamma(\tau) d\tau \right\} \tag{5.44}$$

and

$$\| \mathbf{M} (X_t^s(\omega) - I)^* (X_t^s(\omega) - I) \| \leqslant \left(2 \exp \left\{ \int_s^t \gamma(\tau) d\tau \right\} + 1 \right) \int_s^t \gamma(\tau) d\tau. \tag{5.45}$$

To prove these inequalities, we need a special case of Itô's formula for stochastic operator integrals.

Lemma 1. Let the operator-valued functions $X_s^{(1)}(\omega)$ and $X_s^{(2)}(\omega)$ satisfy the equations

$$X_s^{(i)}(\omega) = X_0^{(i)}(\omega) + \int_0^s A_\tau^{(i)} d\tau + \int_0^s dY_\tau^{(i)}(\omega) B_\tau^{(i)}, \qquad i = 1, 2, \ldots,$$

where $Y_\tau^{(i)}(\omega)$ are Gaussian processes with independent increments, compatible with a filtration $\mathfrak{f}_\tau$, $A_\tau^{(i)}$, $B_\tau^{(i)}$ are $\mathfrak{f}_\tau$-measurable processes such that the following integrals exists:

$$\mathbf{M}\,(dY_\tau^{(1)}(\omega)\,x,\ y)\,(dY_\tau^{(2)}(\omega)\,u,\ v) = \gamma^{1,2}(\tau,\ x,\ y,\ u,\ v)\,d\tau,$$

$x,\ y,\ u,\ v \in \mathbf{X}$. If $x,\ y,\ u,\ v \in \mathbf{X}$, then

$$(X_s^{(1)}(\omega)\,x,\ y)\,(X_s^{(2)}(\omega)\,u,\ v) = (X_0^{(1)}(\omega)\,x,\ y)\,(X_0^{(2)}(\omega)\,u,\ v) +$$

$$+ \int_0^s (X_\tau^{(1)}(\omega)\,x,\ y)\,(A_\tau^{(2)}u,\ v)\,d\tau + \int_0^s (A_\tau^{(1)}x,\ y)\,(X_\tau^{(2)}(\omega)\,u,\ v)\,d\tau +$$

$$+ \int_0^s \gamma^{1,2}(\tau,\ B_\tau^{(1)}x,\ y,\ B_\tau^{(2)}u,\ v)\,d\tau +$$

$$+ \int_0^s (X_\tau^{(1)}(\omega)\,x,\ y)\,(dY_\tau^{(2)}(\omega)\,B_\tau^{(2)}u,\ v) +$$

$$+ \int_0^s (dY_\tau^{(1)}(\omega)\,B_\tau^{(1)}x,\ y)\,(X_\tau^{(2)}(\omega)\,u,\ v). \qquad (5.46)$$

The proof of the lemma is based on Itô's formula for continuous martingales (see [9, p. 96], Theorem 1), since, by the definition of the joint characteristic of two martingales (see (3.36)), the joint characteristic of the continuous martingales $\int_0^s (dY_\tau^{(1)}(\omega)\,B_\tau^{(1)}x,\ y)$ and $\int_0^s (dY_\tau^{(2)}(\omega)\,B_\tau^{(2)}u,\ v)$ is

$$\int_0^s \gamma^{1,2}(\tau,\ B_\tau^{(1)}x,\ y,\ B_\tau^{(2)}u,\ v)\,d\tau.$$

We now return to the proof of inequalities (5.44) and (5.45). By (5.46),

$$\mathbf{M}\,(X_t^s(\omega)\,x,\ e_k)^2 = \mathbf{M}\,(x,\ e_k)^2 +$$

$$+ 2\int_s^t \mathbf{M}\,(X_\tau^s(\omega)\,x,\ e_k)\,(A_\tau X_\tau^s(\omega)\,x,\ e_k)\,d\tau +$$

$$+ \int_s^t \mathbf{M}\beta^t\ (X_\tau^s x,\ e_k,\ X_\tau^s x,\ e_k)\,d\tau.$$

Summing this equality over the basis elements $\{e_k\}$, we obtain

$$\mathbf{M}\,|X_t^s(\omega)x|^2 = |x|^2 + 2\int_s^t \mathbf{M}\,(A_\tau X_\tau^s(\omega)x,\ X_\tau^s(\omega)x) + \int_s^t \mathbf{M}\,(R_\tau X_\tau^s(\omega)x,\ X_\tau^s(\omega)x)\,d\tau \leqslant |x|^2 + \int_s^t \gamma_\tau \mathbf{M}\,|X_\tau^s(\omega)x|^2\,d\tau. \quad (5.47)$$

This implies (5.44). Using (5.47), we find that

$$\mathbf{M}\,|X_t^s(\omega)x - x|^2 = \mathbf{M}\,|X_t^s(\omega)x|^2 + |x|^2 - 2\mathbf{M}\,(x,\ X_t^s(\omega)x) = \mathbf{M}\,|X_t^s(\omega)x|^2 - |x|^2 - 2\mathbf{M}\,(x,\ X_t^s(\omega)x - x) \leqslant \int_s^t \gamma_\tau \mathbf{M}\,|X_\tau^s(\omega)x|^2\,d\tau - 2\int_s^t \mathbf{M}\,(A_\tau X_\tau^s x,\ x)\,d\tau \leqslant \int_s^t [\gamma_\tau \mathbf{M}\,|X_\tau^s(\omega)x|^2 + \|A_\tau\|(\mathbf{M}\,|X_\tau^s(\omega)x|^2 + |x|^2)]\,d\tau.$$

This implies (5.45).

5. For any $x \in \mathbf{X}$, the random function $X_t^s(\omega)\,x$ with values in $\mathbf{X}$ is continuous in t with probability 1. Indeed, $X_t^s(\omega)\,x$ is the sum of a continuous $\mathbf{X}$-valued martingale and the integral

$$\int_s^t A_\tau X_\tau^s(\omega)\,x d\tau. \quad (5.48)$$

Since $|A_\tau X_\tau^s(\omega)\,x| \leqslant \|A_\tau\|\|X_\tau^s(\omega)\,x\|$ and $\mathbf{M}\,|X_\tau^s(\omega)\,x|$ are locally bounded, it follows that $|A_\tau X_\tau^s(\omega)\,x|$ is integrable with probability 1 with respect to τ; hence the integral (5.48) is a continuous function of its upper limit with probability 1.

Another simple linear equation is

$$dZ_t(\omega) = Z_t(\omega)\,A_t dt + Z_t(\omega)\,dY_t(\omega). \quad (5.49)$$

Let $Z_t^s(\omega)$ be the solution of equation (5.49) on $[s, \infty)$ with initial condition $Z_s(\omega) = I$. Arguments analogous to those used to establish properties 1-5 show that they are also valid for $Z_t^s(\omega)$:

(1*) For $0 = t_0 < t_1 < \cdots < t_{n+1}$, the random variables

$\{Z^{t_k}_{t_{k+1}}(\omega), k = 0, \dots, n\}$ are independent.

(2*) For $s \leqslant u \leqslant t$, the following equality holds with probability 1:

$$Z^s_u(\omega)\, Z^u_t(\omega) = Z^s_t(\omega).$$

(3*) The expectation of the solution exists, $\mathbf{M}Z^s_t(\omega) = \tilde{S}^s_t$, for $s \leqslant u \leqslant t$, one has $\tilde{S}^s_t = \tilde{S}^s_u \tilde{S}^u_t$ and

$$\tilde{S}^s_t = I + \int_s^t \tilde{S}^s_\tau A_\tau d\tau.$$

(4*) The estimates (5.44) and (5.45) remain valid if $X^s_t(\omega)$ on the left is replaced by $Z^s_t(\omega)$.

(5*) For any $x \in \mathbf{X}$, the random function $Z^s_t(\omega)$ with values in $\mathbf{X}$ is continuous with respect to t with probability 1.

15.2. Time Reversal in Stochastic Differential Equations

Let us consider equation (5.38) on a finite time interval $t \in [0, \alpha]$. Put

$$\tilde{Y}_t(\omega) = Y_\alpha(\omega) - Y_{\alpha-t}(\omega), \quad Z_t(\omega) = X^\alpha_{\alpha-t}(\omega), \quad \tilde{A}_t = A_{\alpha-t}.$$

Then $Z_t(\omega)$ satisfies the equation

$$dZ_t(\omega) = Z_t(\omega)\tilde{A}_t dt + Z_t(\omega)\, d\tilde{Y}_t(\omega). \tag{5.50}$$

To prove this statement, we first establish the following.

Lemma 2. Let $X_t(\omega)$ be the solution of equation (5.38) with initial condition $X_0(\omega) = I$. Then

$$X_t(\omega) - I =$$

$$= \lim_{n\to\infty} \sum_{k=0}^{n-1} \left(\int_{\frac{kt}{n}}^{\frac{k+1}{n}t} A_\tau d\tau + Y_{\frac{k+1}{n}t}(\omega) - Y_{\frac{k}{n}t}(\omega) \right) X_{\frac{kt}{n}}(\omega)$$

(strong limit in mean square).

Proof. Using the equality

$$X_t(\omega) - I = \sum_{k=0}^{n-1} \left(X_{\frac{k+1}{n}t}(\omega) - X_{\frac{k}{n}t}(\omega)\right)$$

and property 2 $(X_t(\omega) = X_t^0(\omega))$, we obtain

$$X_t(\omega) - I = \sum_{k=0}^{n-1} [X_{\frac{k+1}{n}t}^{\frac{k}{n}t}(\omega) - I]\, X_{\frac{kt}{n}}(\omega).$$

Putting

$$X_{\frac{k+1}{n}t}^{\frac{k}{n}t}(\omega) - I - \int_{\frac{k}{n}t}^{\frac{k+1}{n}t} A_\tau d\tau - \int_{\frac{k}{n}t}^{\frac{k+1}{n}t} dY_\tau(\omega) = U_{n,k}(\omega), \tag{5.51}$$

we have

$$U_{n,k}(\omega) = \int_{\frac{k}{n}t}^{\frac{k+1}{n}t} A_\tau (X_\tau^{\frac{k}{n}t}(\omega) - I)\, d\tau + \int_{\frac{k}{n}t}^{\frac{k+1}{n}t} dY_\tau(\omega) \quad (X_\tau^{\frac{k}{n}t}(\omega) - I).$$

To prove the lemma, it will suffice to show that for all $x \in \mathbf{X}$,

$$\lim_{n\to\infty} \mathbf{M} \left| \sum_{k=0}^{n-1} U_{n,k}(\omega)\, X_{\frac{k}{n}t}(\omega)\, x \right|^2 = 0. \tag{5.52}$$

It follows from the independence of $U_{nk}(\omega)$ and $X_{\frac{k}{n}t}(\omega)$ that

$$\mathbf{M} \,| U_{n,k}(\omega)\, X_{\frac{k}{n}t}(\omega)\, x |^2 =$$

$$= \mathbf{MM}\, [(U_{n,k}^*(\omega)\, U_{n,k}(\omega)\, X_{\frac{k}{n}t}(\omega)\, x, \quad X_{\frac{k}{n}t}(\omega)\, x) / X_{\frac{k}{n}t}(\omega)\, x] =$$

$$= \mathbf{M}\, (\{\mathbf{M} U_{n,k}^*(\omega)\, U_{n,k}(\omega)\}\, X_{\frac{k}{n}t}(\omega)\, x, \quad X_{\frac{k}{n}t}(\omega)\, x).$$

Using inequalities (5.44) and (5.45), we can show that for some $\delta > 0$

$$\| \mathbf{M} U_{n,k}^*(\omega)\, U_{n,k}(\omega) \| \leqslant \frac{\delta}{n^2}. \tag{5.53}$$

Therefore $\mathbf{M}\, | U_{n,k}(\omega)\, X_{\frac{k}{n}t}(\omega)\, x |^2 \leqslant \frac{\delta}{n^2} \mathbf{M}\, | X_{\frac{k}{n}t}(\omega)\, x |^2 \leqslant \frac{\delta_1}{n^2} | x |^2$,
where δ_1 is a constant (this is proved using inequality (5.44)). The last estimate implies (5.52) and the proof of the lemma

is thus complete. ■

Remark. Using the representation

$$X_t(\omega) - I = \sum_{k=0}^{n-1} [X_t^{\frac{k}{n}t}(\omega) - X_t^{\frac{k+1}{n}t}(\omega)] =$$

$$= \sum_{k=0}^{n-1} X_t^{\frac{k+1}{n}t}(\omega)\,[X^{\frac{k}{n}t}_{\frac{k+1}{n}t}(\omega) - I],$$

and the relationships (5.51), (5.53) and proceeding as in Lemma 2, we can prove the equality

$$X_t(\omega) - I = \lim_{n\to\infty} \sum_{k=0}^{n-1} X_t^{\frac{k+1}{n}t}(\omega)\left(\int_{\frac{kt}{n}}^{\frac{k+1}{n}t} A_\tau d\tau + Y_{\frac{k+1}{n}t}(\omega) - Y_{\frac{k}{n}t}(\omega)\right). \quad (5.54)$$

Using (5.54), we can write

$$X_\alpha^{\alpha-t}(\omega) - I = \lim_{n\to\infty} \sum_{k=0}^{n-1} X_\alpha^{\alpha-\frac{k}{n}t}(\omega)\left(\int_{\alpha-\frac{k+1}{n}t}^{\alpha-\frac{k}{n}t} A_\tau d\tau + Y_{\alpha-\frac{k}{n}t}(\omega) - Y_{\alpha-\frac{k+1}{n}t}(\omega)\right)$$

(equality (5.54) is applied to the process $V_s(\omega) = X_{\alpha-t+s}^{\alpha-t}(\omega)$ for $s = t$). Replacing A by $\tilde{A}$, Y by $\tilde{Y}$ and X by Z in the last equality, we obtain

$$Z_t(\omega) - I = \lim_{n\to\infty} \sum_{k=0}^{n-1} Z_{\frac{k}{n}t}(\omega)\left(\int_{\frac{k}{n}t}^{\frac{k+1}{n}} \tilde{A}_\sigma d\sigma + \tilde{Y}_{\frac{k+1}{n}t}(\omega) - \tilde{Y}_{\frac{k}{n}t}(\omega)\right). \quad (5.55)$$

Therefore,

$$Z_t(\omega) - I = \int_0^t Z_\sigma(\omega)\tilde{A}_\sigma d\sigma + \int_0^t Z_\sigma(\omega)\, d\tilde{Y}_\sigma(\omega) +$$

$$+ \lim_{n\to\infty}\left(\sum_{k=0}^{n-1}\int_{\frac{k}{n}t}^{\frac{k+1}{n}t} [Z_\sigma(\omega) - Z_{\frac{k}{n}t}(\omega)]\tilde{A}_\sigma d\sigma + \right.$$

$$\left. + \int_{\frac{k}{n}t}^{\frac{k+1}{n}t} [Z_\sigma(\omega) - Z_{\frac{k}{n}t}(\omega)]\, d\tilde{Y}_\sigma(\omega)\right).$$

To prove (5.50), it now remains to prove that the limit on the right (which exists in the sense of strong convergence in mean square) is equal to zero. We have the estimate

$$\mathbf{M}\left|\int_{\frac{k}{n}t}^{\frac{k+1}{n}t} (Z_\sigma(\omega) - Z_{\frac{k}{n}t}(\omega))\tilde{A}_\sigma x\, d\sigma\right|^2 \leqslant$$

$$\leqslant \frac{t}{n}\int_{\frac{k}{n}t}^{\frac{k+1}{n}t} \mathbf{M}\,|(Z_\sigma(\omega) - Z_{\frac{k}{n}t}(\omega))\tilde{A}_\sigma x|^2\, d\sigma \leqslant$$

$$\leqslant \frac{t}{n}\delta|x|^2 \int_{\frac{k}{n}t}^{\frac{k+1}{n}t} \|\mathbf{M}(Z_\sigma(\omega) - Z_{\frac{k}{n}t}(\omega))^*(Z_\sigma(\omega) -$$

$$- Z_{\frac{k}{n}}(\omega))\|\, d\sigma = O\left(\frac{1}{n^3}\right)$$

(this is proved using (5.45)). Consequently,

$$\mathbf{M}\left|\left\{\sum_{k=0}^{n-1}\int_{\frac{k}{n}t}^{\frac{k+1}{n}t} (Z_\sigma(\omega) - Z_{\frac{k}{n}t}(\omega))\tilde{A}_\sigma d\sigma\right\} x\right|^2 = nO\left(\frac{1}{n^3}\right) \to 0.$$

Using the fact that the terms in the sum

$$Z_n(\omega) = \sum_{k=0}^{n-1} \int_{\frac{k}{n}t}^{\frac{k+1}{n}t} [Z_\sigma(\omega) - Z_{\frac{k}{n}t}(\omega)]\, d\tilde{Y}_\sigma(\omega)\, x$$

are uncorrelated, we obtain

$$\mathbf{M}\,|Z_n(\omega)|^2 = \sum_{k=0}^{n-1} \mathbf{M} \left| \int_{\frac{k}{n}t}^{\frac{k+1}{n}t} [Z_\sigma(\omega) - Z_{\frac{k}{n}t}(\omega)]\, d\tilde{V}_\sigma(\omega)\, x \right|^2 =$$

$$= O\left(\sum_{k=0}^{n-1} \int_{\frac{k}{n}}^{\frac{k+1}{n}t} \|\mathbf{M}\,(Z_\sigma(\omega) - Z_{\frac{k}{n}t}(\omega))^* (Z_\sigma(\omega) -$$

$$- Z_{\frac{k}{n}t}(\omega))\| d\sigma\right) = nO\left(\frac{1}{n^3}\right) \to 0$$

(here we have used inequality (5.44)). This completes the proof of equation (5.50).

15.3. Definition of Stochastic Semigroups

Properties 1 and 2 (or 2*) of the solutions of elementary linear equations are basic for the stochastic semigroups to be defined now. A family of operators $X_t^s(\omega)$ with values in $\mathbf{L}_s(\Omega, \mathbf{X})$, defined for $0 \leqslant s \leqslant t \leqslant T$ (or $0 \leqslant s < t < \infty$), is called a (strong) stochastic semigroup if the following conditions hold:

I. Let $\mathfrak{f}_t^s$ be the σ-algebra generated by $\{X_\tau^\sigma(\omega)\,x,\ x \in \mathbf{X},\ s \leqslant \sigma \leqslant \tau \leqslant t\}$. Then the σ-algebras $\mathfrak{f}_t^0$ and $\mathfrak{f}_T^t$, $0 < t < T$, are independent.

II. For $u \leqslant s \leqslant t$,

$$X_t^s(\omega)\, X_s^u(\omega) = X_t^u(\omega), \qquad X_t^t(\omega) = I.$$

II*. For $u \leqslant s \leqslant t$,

$$X_s^u(\omega)\, X_t^s(\omega) = X_t^u(\omega), \quad X_t^t(\omega) = I.$$

In case II (II*) the semigroup is called a left (right) semigroup. It is easy to see that if $X_t^s(\omega)$ is a left semigroup,

$0 \leqslant s \leqslant t \leqslant T$, then $\hat{X}_t^s(\omega) = X_{T-s}^{T-t}(\omega)$ is a right semigroup. Since we are studying semigroups on finite intervals, we may restrict attention to one or the other of the two types, say only left semigroups. Semigroups which are solutions of linear equations of type (5.38) possess certain additional regularity properties:

III. $\forall x \in \mathbf{X}, X_t^s(\omega)x$ is strongly continuous in t;

IV. $\|\mathbf{M}(X_t^s(\omega))^* X_t^s(\omega)\|$ is locally bounded and

$$\lim_{\substack{t-s\to 0\\ t,s\leqslant T}} \|\mathbf{M}(X_t^s(\omega)-I)^*(X_t^s(\omega)-I)\| = 0.$$

The strong continuity of the solution is established, e.g., in Theorem 1 of § 12. Condition IV follows from (5.44) and (5.45).

We are now going to study the structure of semigroups that satisfy conditions I-IV.

Denote $Q_t^s = \mathbf{M}X_t^s(\omega)$. Existence, and hence also boundedness, of the expectation follow from condition IV. From condition II we deduce (in view of the independence of $X_t^s(\omega)$ and $X_s^u(\omega)$, which is a corollary of condition I) that for $u \leqslant s \leqslant t$

$$Q_t^u = \mathbf{M}X_t^s(\omega)X_s^u(\omega) = \mathbf{M}X_t^s(\omega)\mathbf{M}X_s^u(\omega) = Q_t^sQ_s^u. \qquad (5.56)$$

In addition,

$$\|Q_t^s - I\|^2 = \sup_{|x|\leqslant 1} |\mathbf{M}(X_t^s(\omega)x - x)|^2 \leqslant \sup_{|x|\leqslant 1} \mathbf{M}|X_t^s(\omega)x - x|^2 =$$

$$= \sup_{|x|\leqslant 1} (\mathbf{M}(X_t^s(\omega)-I)^*(X_t^s(\omega)-I)x, x) =$$

$$= \|\mathbf{M}(X_t^s(\omega)-I)^*(X_t^s(\omega)-I)\|.$$

Thus

$$\lim_{\substack{t-s\to 0\\ t,s\leqslant T}} \|Q_t^s - I\| = 0. \qquad (5.57)$$

It follows from (5.56) that the function $Q_t = Q_t^0$ with values in $\mathbf{L}(\mathbf{X})$ is continuous in norm and, for all $t > 0$ the function Q_t^{-1} is also defined and continuous in t (since for $0 = t_0 < t_1 < \ldots < t_n = t$, we have $Q_t = Q_t^{t_{n-1}} \ldots Q_{t_1}$ and for sufficiently small $\max\limits_k (t_{k+1} - t_k)$ all factors in the product are invertible thanks to (5.57)).

Now put

$$\hat{X}_t^s(\omega) = Q_t^{-1} X_t^s(\omega) Q_s. \tag{5.58}$$

We claim that $\hat{X}_t^s(\omega)$ is also a (left) semigroup satisfying conditions III and IV. It is easy to see that the σ-algebra generated by $\{\hat{X}_\tau^\sigma(\omega),\ x \in \mathbf{X},\ s \leqslant \sigma < \tau < t\}$ is exactly $\mathfrak{f}_t^s$, so that condition I holds. Next, for $u \leqslant s \leqslant t$

$$\hat{X}_t^s(\omega)\hat{X}_s^u(\omega) = Q_t^{-1} X_t^s(\omega) Q_s Q_s^{-1} X_s^u(\omega) Q_u =$$

$$Q_t^{-1} X_t^s(\omega) X_s^u(\omega) Q_u = \hat{X}_t^u(\omega)$$

proving II. Now, by condition III, $X_t^s(\omega)\ Q_s\ x$ is strongly continuous in t, and Q_t^{-1} is continuous in the $\mathbf{L}(\mathbf{X})$-norm; hence the function

$$Q_t^{-1} X_t^s(\omega) Q_s x$$

is also continuous in t. Finally,

$$\|\mathbf{M}(\hat{X}_t^s(\omega) - I)^*(\hat{X}_t^s(\omega) - I)\| = \sup_{|x| \leqslant 1} \mathbf{M}\,|\hat{X}_t^s(\omega)x - x|^2 \leqslant$$

$$\leqslant 2 \sup_{|x| \leqslant 1} (\mathbf{M}\,|Q_t^{-1}(X_t^s(\omega) - I) Q_s x|^2 + |Q_t^{-1}(Q_s - Q_t)x|^2) \leqslant$$

$$\leqslant 2\|Q_t^{-1}\|^2 (\|Q_s\|^2 \|\mathbf{M}(X_t^s(\omega) - I)^*(X_t^s(\omega) - I)\| + \|Q_s - Q_t\|^2).$$

The right-hand side of this equality is bounded and tends to zero, by condition IV and the continuity of Q_t.

Note that

$$\mathbf{M}\hat{X}_t^s(\omega) = Q_t^{-1}\mathbf{M}X_t^s(\omega) Q_s = Q_t^{-1} Q_t^s Q_s = Q_t^{-1} Q_t = I.$$

Hence $\hat{X}_t^s(\omega)$ satisfies the following conditions:

V. $\hat{X}_t^s(\omega)$ is a martingale (as a function of t) adapted to $\mathfrak{f}_t^s$. Indeed, for $s < t_1 < t_2$,

$$\mathbf{M}\,[(\hat{X}_{t_2}^s(\omega)x, y)/\mathfrak{f}_{t_1}^s] = \mathbf{M}\,[(\hat{X}_{t_2}^{t_1}(\omega)\hat{X}_{t_1}^s(\omega)x, y)/\mathfrak{f}_{t_1}^s] =$$

$$= ([\mathbf{M}\hat{X}_{t_2}^{t_1}(\omega)]\hat{X}_{t_1}^s(\omega)x, y) = (\hat{X}_{t_1}^s(\omega)x, y)$$

(we have used the fact that $\hat{X}_{t_2}^{t_1}(\omega)$ is independent of $\mathfrak{f}_{t_1}^s$ and the $\mathfrak{f}_{t_1}^s$-measurability of $\hat{X}_{t_1}^s(\omega)$. Thus, the investigation of semigroups satisfying conditions I-IV may be reduced via the transformation (5.58) to the investigation of semigroups that also satisfy the additional condition V. Moreover, if $\hat{X}_t^s(\omega)$ satisfies conditions I-V, then, for any ivertible function Q_t which is continuous in norm,

$$X_t^s(\omega) = Q_t \hat{X}_t^s(\omega) Q_s^{-1}$$

is a semigroup satisfying conditions I-IV.

15.4. Semigroups which are Martingales

In this subsection we study semigroups $X_t^s(\omega)$, that satisfy conditions I-V. Put

$$R_t^s = \mathbf{M}(X_t^s(\omega))^* X_t^s(\omega).$$

Let $u < s < t$. Then, since $X_s^u(\omega)$ and $X_t^s(\omega)$ are independent and $\mathbf{M}X_t^s(\omega) = I$ we have

$$R_t^u = \mathbf{M}(X_s^u(\omega))^* (X_t^s(\omega))^* X_t^s(\omega) X_s^u(\omega) =$$
$$= \mathbf{M}(X_s^u(\omega))^* R_t^s X_s^u(\omega) = R_t^s + \mathbf{M}(X_s^u(\omega) - I)^* R_t^s (X_s^u(\omega) -$$
$$- I) = R_s^u + \mathbf{M}(X_s^u(\omega))^* [R_t^s - I] X_s^u(\omega).$$

In particular,

$$R_t^s = I + \mathbf{M}(X_t^s(\omega) - I)^* (X_t^s(\omega) - I).$$

Since $\mathbf{M}(X_t^s(\omega) - I)^* (X_t^s(\omega) - I)$ is a nonnegative symmetric operator, as is R_t^s, it follows that $R_t^s - I \geqslant 0$ and, for $s \leqslant \sigma < \tau \leqslant t$ and all $y \in \mathbf{X}$,

$$(R_t^s y, y) \geqslant (R_\tau^\sigma y, y) \geqslant (y, y). \tag{5.59}$$

Lemma 3. Let $s = t_0 < t_1 < \ldots < t_n = t$. Then

$$\left\| \mathbf{M} \sum_{k=0}^{n-1} (X_{t_{k+1}}^{t_k}(\omega) - I)^* (X_{t_{k+1}}^{t_k}(\omega) - I) \right\| \leqslant \| R_t^s - I \|. \tag{5.60}$$

Proof. We have

$$X_t^s(\omega) - I = \sum_{k=0}^{n-1} (X_t^{t_k}(\omega) - X_t^{t_{k+1}}(\omega)) =$$

$$= \sum_{k=0}^{n-1} X_t^{t_{k+1}}(\omega)\,(X_{t_{k+1}}^{t_k}(\omega) - I). \tag{5.61}$$

For $i < k$,

$$\mathbf{M}\,(X_t^{t_{i+1}}(\omega)\,[X_{t_{i+1}}^{t_i}(\omega) - I]\,x,\ X_t^{t_{k+1}}(\omega)\,[X_{t_{k+1}}^{t_k}(\omega) - I]\,x) =$$

$$\mathbf{M}\,(X_t^{t_{i+1}}(\omega)\,\mathbf{M}\,[X_{t_{i+1}}^{t_i}(\omega) - I]\,x,\ X_t^{t_{k+1}}(\omega)\,[X_{t_{k+1}}^{t_k}(\omega) - I]\,x) = 0 \tag{5.62}$$

(we have used the fact that $X_{t_{i+1}}^{t_i}(\omega)$ is independent of the σ-algebra $\mathfrak{f}_t^{t_{i+1}}$, relative to which the other operators in the last equality are measurable. Therefore,

$$\mathbf{M}\,|(X_t^s(\omega) - I)\,x|^2 = \sum_{k=0}^{n-1} \mathbf{M}\,|X_t^{t_{k+1}}(\omega)\,(X_{t_{k+1}}^{t_k}(\omega) - I)\,x|^2.$$

Since

$$\mathbf{M}\,|X_t^{t_{k+1}}(\omega)\,(X_{t_{k+1}}^{t_k}(\omega) - I)\,x|^2 =$$

$$= \mathbf{M}\mathbf{M}\,(|X_t^{t_{k+1}}(\omega)\,(X_{t_{k+1}}^{t_k}(\omega) - I)\,x|^2/\mathfrak{f}_{t_{k+1}}^{t_k}) =$$

$$= \mathbf{M}\,(R_t^{t_{k+1}}\,(X_{t_{k+1}}^{t_k}(\omega) - I)\,x,\ (X_{t_{k+1}}^{t_k}(\omega) - I)\,x) \geqslant$$

$$\geqslant \mathbf{M}\,|(X_{t_{k+1}}^{t_k}(\omega) - I)\,x|^2$$

(we have used (5.59)), it follows that

$$\mathbf{M}\,|(X_t^s(\omega) - I)\,x|^2 \geqslant \sum_{k=0}^{n-1} \mathbf{M}\,|(X_{t_{k+1}}^{t_k}(\omega) - I)\,x|^2.$$

Evaluating the supremum over $|x| \leqslant 1$, we obtain (5.60). ■

Remark. Assume that, besides conditions I and II, the semigroup $X_t^s(\omega)$ also satisfies the following condition: the distribution of $X_{t+h}^t(\omega)$ depends only on h; a semigroup satisfying this condition is said to be homogeneous. Given a homogeneous subgroup such that, for some t,

$$\|\mathbf{M}\,(X_t^0(\omega))^*\,(X_t^0(\omega))\| < \infty$$

and condition V holds, then condition IV also holds. Indeed, it follows from the proof of Lemma 3 that

$$\mathbf{M}\,|X_t^0(\omega)x - x|^2 \geqslant n\mathbf{M}\,|(X^0_{\frac{t}{n}}(\omega) - I)x|^2.$$

Therefore,

$$\|\mathbf{M}(X^0_{\frac{t}{n}}(\omega) - I)^*(X^0_{\frac{t}{n}}(\omega) - I)\| \leqslant \frac{1}{n}\|\mathbf{M}(X_t^0(\omega) - I)^* \quad (X_t^0(\omega) - I)\|.$$

Remark 2. Using the representation (5.61), and the notation of the lemma, we have

$$X_t^s(\omega) - I - \sum_{k=0}^{n-1}(X_{t_{k+1}}^{t_k}(\omega) - I) =$$
$$= \sum_{k=0}^{n-1}(X_t^{t_{k+1}}(\omega) - I)(X_{t_{k+1}}^{t_k}(\omega) - I).$$

Hence, using (5.62), we find that for $x \in \mathbf{X}$

$$\mathbf{M}\left|\left[X_t^s(\omega) - I - \sum_{k=0}^{n-1}(X_{t_{k+1}}^{t_k}(\omega) - I)\right]x, x\right|^2 \leqslant \tag{5.63}$$
$$\leqslant \|R_t^s - I\| \sum_{k=0}^{n-1}\mathbf{M}\,|(X_{t_{k+1}}^{t_k}(\omega) - I)x|^2.$$

Corollary. Consider two partitions of the interval $[s, t]$, one a sub-partition of the other: $s = t_0 < \ldots < t_n = t$ and $s = \tau_0 < \tau_1 < \ldots < \tau_m = t$, $n < m$. Then

$$\mathbf{M}\left|\left[\sum_{k=0}^{n-1}(X_{t_{k+1}}^{t_k}(\omega) - I) - \sum_{j=0}^{m-1}(X_{\tau_{j+1}}^{\tau_j}(\omega) - I)\right]x\right|^2 \leqslant \tag{5.64}$$
$$\leqslant \max_k \|R_{t_{k+1}}^{t_k} - I\| \sum_{j=0}^{m-1}\mathbf{M}\,|(X_{t_{j+1}}^{t_j}(\omega) - I)x|^2.$$

To prove this, we use the fact that the terms are uncorrelated, writing

$$\mathbf{M}\left|\left[\sum_{k=0}^{n-1}(X_{t_{k+1}}^{t_k}(\omega) - I) - \sum_{j=0}^{m-1}(X_{\tau_{j+1}}^{\tau_j}(\omega) - I)\right]x\right|^2 =$$
$$= \sum_{k=0}^{n-1}\mathbf{M}\left|\left[X_{t_{k+1}}^{t_k}(\omega) - I - \sum_{t_k \leqslant \tau_j < t_{k+1}}(X_{\tau_{j+1}}^{\tau_j}(\omega) - I)\right]x\right|^2,$$

then applying inequality (5.63) to each term of the sum

$\sum_{k=0}^{n-1}$ on the right.

Theorem 1. If $X_t^s(\omega)$, $0 \leqslant s < t \leqslant T$, satisfies conditions I-V, then there exists a strongly continuous Gaussian process $Y_t(\omega)$ on $[0, T]$, with independent increments and values in $\mathbf{L}_s(\Omega, \mathbf{X})$, such that $\mathbf{M}Y_t(\omega) = 0$ and

$$\lim_{t-s\to 0} \|\mathbf{M}(Y_t(\omega) - Y_s(\omega))^*(Y_t(\omega) - Y_s(\omega))\| = 0 \tag{5.65}$$

such that for all $s < t$,

$$Y_t(\omega) - Y_s(\omega) = \lim_{\max\Delta t_k\to 0} \sum (X_{t_{k+1}}^{t_k}(\omega) - I) \tag{5.66}$$

where the limit in (5.56) is in the sense of strong convergence in mean square of operators in $\mathbf{L}_s(\Omega, \mathbf{X})$, $s = t_0 < \cdots < t_n = t$, $t_{k+1} - t_k = \Delta t_k$.

Proof. Let $s = s_0 < \cdots, < s_m = t$, $s = t_0 < \cdots < t_n = t$ be two partitions, and denote their intersection by $s = \tau_0 < \cdots < \tau_q = t$. If $\max \Delta s_j < \delta$ and $\max \Delta t_k < \delta$, then, by (5.64) and (5.60), estimating the difference between the sums based on the first and second partition and the sum based on the third, we obtain

$$\mathbf{M}\left|\left[\sum_{k=0}^{n-1}(X_{t_{k+1}}^{t_k}(\omega) - I) - \sum_{j=0}^{m-1}(X_{s_{j+1}}^{s_j}(\omega) - I)\right]x\right|^2 \leqslant$$

$$\leqslant 2\max_t \|R_{t+\delta}^t - I\|\|R_t^s - I\|.$$

Therefore the limit on the right of (5.65) exists. Denote this limit by $Y_t^s(\omega)$. Since $Y_t^s(\omega)$ is $\mathfrak{f}_t^s$-measurable, it follows that $Y_t^s(\omega)$ is independent of $\mathfrak{f}_s^0$. $\mathbf{M}Y_t^s(\omega) = 0$, since the r.v. whose limit appears on the right of (5.66) has expectation zero. Setting $Y_t^0(\omega) = Y_t(\omega)$, we have $Y_t^s(\omega) = Y_t(\omega) - Y_s(\omega)$. From (5.63), in view of (5.66), we deduce that

$$\mathbf{M}|(X_t^s(\omega) - I - Y_t(\omega) + Y_s(\omega))x|^2 \leqslant \tag{5.67}$$

$$\leqslant \|R_t^s - I\|((R_t^s - I)x, x),$$

whence follows (5.65). $Y_t(\omega)$ is a process with independent increments. If we can prove that it is strongly continuous, this will also imply that it is Gaussian, completing the proof of our theorem.

Choose a sequence of partitions of the interval $[0, T]$: $0 = t_{n0} < t_{n1} < \dots < t_{nn} = T$, such that $\lim_{n\to\infty} \max_k (t_{nk+1} - t_{nk}) = 0$.

Construct the sequence of continuous operator-valued martingales

$$X_t^{(n)}(\omega) = \sum_{k=0}^{j-1} [X_{t_{nk+1}}^{t_{nk}}(\omega) - I] + X_t^{t_{nj}}(\omega) - I, \quad t \in [t_{nj}, t_{nj+1}].$$

These martingales, like the martingale $Y_t(\omega)$, are adapted to the filtration $\mathfrak{f}_t$. Since, as proved, $X_t^{(n)}(\omega)$ is strongly convergent in mean square to $Y_t(\omega)$, it follows that $Y_t(\omega)$ is also a continuous martingale (see Theorem 5, § 9).■

The process $Y_t(\omega)$, whose existence is established in Theorem 1 will be called the Gaussian process associated with the semigroup $X_t^s(\omega)$. It turns out that, given $Y_t(\omega)$, the whole semigroup $X_t^s(\omega)$ is uniquely determined.

Theorem 2. Let $Y_t(\omega)$ be the process associated with the semigroup $X_t^s(\omega)$. Then

$$X_t^s(\omega) - I = \lim_{\max\Delta t_k \to 0} \sum_{k=0}^{n-1} X_t^{t_{k+1}}(\omega)[Y_{t_{k+1}}(\omega) - Y_{t_k}(\omega)]; \tag{5.68}$$

$$X_t^s(\omega) = \lim_{\max\Delta t_k \to 0} (I + Y_{t_n}(\omega) - Y_{t_{n-1}}(\omega))(I + Y_{t_{n-1}}(\omega) - Y_{t_{n-2}}(\omega)) \dots (I + Y_{t_1}(\omega) - Y_{t_0}(\omega)), \tag{5.69}$$

where both limits are in the sense of strong convergence, $s = t_0 < t_1 < \dots < t_n = t$, $\Delta t_k = t_{k+1} - t_k$.

Proof. Using the representation (5.61), we write

$$X_t^s(\omega) - I - \sum_{k=0}^{n-1} [Y_{t_{k+1}}(\omega) - Y_{t_k}(\omega)] X_{t_k}^s(\omega) =$$
$$= \sum_{k=0}^{n-1} X_t^{t_{k+1}}(\omega)[X_{t_{k+1}}^{t_k}(\omega) - I - Y_{t_{k+1}}(\omega) + Y_{t_k}(\omega)].$$

Consequently, on the basis of (5.67), using Lemma 3, we obtain

$$\mathbf{M}\left|\sum_{k=0}^{n-1} X_t^{t_{k+1}}(\omega)[X_{t_{k+1}}^{t_k}(\omega) - I - Y_{t_{k+1}}(\omega) + Y_{t_k}(\omega)]\,x\right|^2 \leqslant$$
$$\leqslant \sum_{k=0}^{n-1} \| R_t^{t_{k+1}} \| \| R_{t_{k+1}}^{t_k} - I \| ((R_{t_{k+1}}^{t_k} - I)\,x, x) \leqslant$$
$$\leqslant \| R_t^s \| \| R_t^s - I \| \max_k \| R_{t_{k+1}}^{t_k} - I \| |x|^2.$$

This proves (5.68).

Denote $(i < j)$

$$Z_j^i(\omega) = (I + Y_{t_j}(\omega) - Y_{t_{j-1}}(\omega)) \ \ldots \ (I + Y_{t_{i+1}}(\omega) - Y_{t_i}(\omega)).$$

Then

$$Z_n^k(\omega) - I = \sum_{j=k+1}^{n} Z_n^j(\omega)\,[Z_j^{j-1}(\omega) - I].$$

Therefore, using the independence of $Z_n^j(\omega)$ and $Y_{t_j}(\omega) - Y_{t_{j-1}}(\omega)$, we get

$$\mathbf{M}\,|Z_n^k(\omega)\,x|^2 = (x, x) + \sum_{j=k+1}^{n} \mathbf{M}\,|Z_n^j(\omega)\,[Y_{t_j}(\omega) -$$

$$- Y_{t_{j-1}}(\omega)]\,x|^2 \leqslant (x, x) + \sum_{j=k+1}^{n} \|\mathbf{M}\,(Z_n^j(\omega))^* \times$$

$$\times (Z_n^j(\omega))\|\,(\mathbf{M}\,(Y_{t_j}(\omega) - Y_{t_{j-1}}(\omega))^*\,(Y_{t_j}(\omega) - Y_{t_{j-1}}(\omega))\,x,\ x) \leqslant$$

$$\leqslant (x, x) + \sup_{k \leqslant j \leqslant n} \|\mathbf{M}\,(Z_n^j(\omega))^*\,(Z_n^j(\omega))\| \times$$

$$\times\, \mathbf{M}\,(Y_{t_n}(\omega) - Y_{t_k}(\omega))^*\,(Y_{t_n}(\omega) - Y_{t_k}(\omega))\,\|\,(x, x).$$

If $s < t$ are such that

$$\|\mathbf{M}\,(Y_t(\omega) - Y_s(\omega))^*\,(Y_t(\omega) - Y_s(\omega))\| \leqslant \gamma < 1, \qquad (5.70)$$

then

$$\sup_k \sup_{|x| \leqslant 1} |Z_n^k(\omega)\,x|^2 \leqslant \gamma \sup_j \|\mathbf{M}\,(Z_n^j(\omega))^*\,Z_n^j(\omega)\| + 1$$

and so

$$\|\mathbf{M}\,(Z_n^k(\omega))^*\,Z_n^k(\omega)\| \leqslant \frac{1}{1-\gamma}.$$

It is obviously sufficient to prove formula (5.69) for sufficiently small $t - s$. Suppose that (5.70) holds with $\gamma < \frac{1}{2}$; then

$$Z_n^k(\omega) - X_t^{t_k}(\omega) = \sum_{j=k+1}^{n} Z_n^j(\omega)[Z_j^{j-1}(\omega) - I] -$$

$$- \sum_{j=k+1}^{n} X_t^{t_j}(\omega)[X_{t_j}^{t_{j-1}}(\omega) - I] =$$

$$= \sum_{j=k+1}^{n} [Z_n^j(\omega) - X_t^{t_j}(\omega)][Z_j^{j-1}(\omega) - I] +$$

$$+ \sum_{j=k+1}^{n} X_t^{t_j}(\omega)[Z_j^{j-1}(\omega) - X_{t_j}^{t_{j-1}}(\omega)].$$

Again using the fact that the terms in any sum are uncorrelated (5.70), we obtain

$$\mathbf{M}|(Z_n^k(\omega) - X_t^{t_k}(\omega))x|^2 \leqslant 2 \sum_{j=k+1}^{n} \mathbf{M}|[Z_n^j(\omega) -$$

$$- X_t^{t_j}(\omega)][Y_{t_j}(\omega) - Y_{t_{j-1}}(\omega)]x|^2 +$$

$$+ 2 \sum_{j=k+1}^{n} \mathbf{M}|X_t^{t_j}(\omega)[I + Y_{t_j}(\omega) - Y_{t_{j-1}}(\omega) - X_{t_j}^{t_{j-1}}(\omega)]x|^2 \leqslant$$

$$\leqslant 2 \sum_{j=k+1}^{n} \|\mathbf{M}(Z_n^j(\omega) - X_t^{t_j}(\omega))^*(Z_n^j(\omega) - X_t^{t_j}(\omega))\| \times$$

$$\times (\mathbf{M}(Y_{t_j}(\omega) - Y_{t_{j-1}}(\omega))^*(Y_{t_j}(\omega) - Y_{t_{j-1}}(\omega))x, x) +$$

$$+ 2 \sum_{j=k+1}^{n} \|R_t^{t_j}\| \|R_{t_j}^{t_{j-1}} - I\| ((R_{t_j}^{t_{j-1}} - I)x, x).$$

Hence the following inequality:

$$\sup_{|x|\leqslant 1} \sup_k \mathbf{M}|(Z_n^k(\omega) - X_t^{t_k}(\omega))x|^2 \leqslant$$

$$\leqslant 2\gamma \sup \|\mathbf{M}[Z_n^j(\omega) - X_t^{t_j}(\omega)]^*[Z_n^j(\omega) - X_t^{t_j}(\omega)]\| +$$

$$+ 2\|R_t^s\| \max_j \|R_{t_j}^{t_{j-1}} - I\| \|R_t^s - I\|.$$

Consequently,

$$\|\mathbf{M}(Z_n^0(\omega) - X_t^s(\omega))^*(Z_n^0(\omega) - X_t^s(\omega))\| \leqslant$$

$$\leqslant 2\|R_t^s\| \|R_t^s - I\| \frac{\max_j \|R_{t_j}^{t_{j-1}} - I\|}{1 - 2\gamma}.$$

Letting $\Delta t_j \to 0$, we obtain (5.69).■

Remark. Using time reversal, we can write (5.68) as

$$\tilde{X}_t^s(\omega) = I + \lim_{\max \Delta t_k \to 0} \sum_{k=0}^{n-1} \tilde{X}_{t_k}^s(\omega)\,[\tilde{Y}_{t_{k+1}}(\omega) - \tilde{Y}_{t_k}(\omega)], \qquad (5.71)$$

where

$$\tilde{X}_t^s(\omega) = X_{T-s}^{T-t}, \quad \tilde{Y}_t(\omega) = Y_T(\omega) - Y_{T-t}(\omega),$$

$$s = t_0 < t_1 < \cdots < t_n = t.$$

Since $\tilde{X}_t^s(\omega)$ is continuous in mean square as a function of t, it follows that the limit on the right of (5.71) coincides with the stochastic integral, and so $\tilde{X}_t^s(\omega)$, as a function of t, satisfies the equation

$$X_t^s(\omega) = I + \int_s \tilde{X}_t^s(\omega)\, d\tilde{Y}_t(\omega). \qquad (5.72)$$

BIBLIOGRAPHY

1. Akhiezer, N. I., Glazman, I. M. Theory of Linear Operators in Hilbert Space. Moscow, 'Nauka', 1966. 543 pp. [Russian].

2. Belopol'skaya, Ya. I., Daletskii, Yu. L. 'Diffusion Processes in Smooth Banach Manifolds', *Trudy Mosk. Mat. Obshch.*, 1977, 37. [Russian].

3. Butsan, G. P. *Stochastic Semigroups*. Kiev, 'Naukova Dumka', 1977. 216 pp. [Russian].

4. Bakhaniya, N. N. *Probability Distributions in Linear Spaces*. Tbilisi, 'Metsniereba', 1971. 153 pp. [Russian].

5. Vershik, A. M., Sudakov, V. N. 'Probability Measures in Infinite-dimensional Spaces', *Zap. Nauchn. Seminara Leningrad. Otdeleniya Mat. Inst.*, 1969, 12 pp. 7-67. [Russian].

6. Gikhman, I. I., Skorokhod, A. V. *Introduction to the Theory of Random Processes*. Moscow, 'Nauka', 1965. 632 pp. [Russian].

7. Gikhman, I. I., Skorokhod, A. V. *Stochastic Differential Equations*. Kiev, 'Naukova Dumka', 1968. 354 pp. [Russian].

8. Gikhman, I. I., Skorokhod, A. V. *Theory of Random Processes*, I. Moscow, 'Nauka', 1971. 664 pp. [Russian].

9. Gikhman, I. I., Skorokhod, A. V. *Theory of Random Processes*, III. Moscow, 'Nauka', 1975. 496 pp. [Russian].

10. Girko, V. L. *Random Matrices*. Kiev, 'Vishcha Shkola', 1975. 447 pp. [Russian].

11. Grenander, U. *Probabilities on Algebraic Structures*, New York, Wiley, 1963.

12. Daletskii, Yu. L. 'Infinite-dimensional Elliptic Operators and the Associated Parabolic Equations', *Uspekhi Mat. Nauk*, 1967, 22, no. 4, pp. 3-54. [Russian].

13. Kolmogorov, A. N. *Basic Concepts of Probability Theory*. Moscow-Leningrad, ONTI, 1936. 79 pp. [Russian].

14. Marchenko, V. A., Pastur, L. A. 'Distribution of Eigenvalues in Certain Ensembles of Random Matrices', *Mat. Sbornik*, 1967, 72, no. 4, pp. 507-536. [Russian].

15. Minlos, R. A. 'Generalized Random Processes and their Extension to Measures', *Trudy Mosk. Mat. Obshch.*, 1958, 8, pp. 497-518. [Russian].

16. Pastur, L. A. 'Spectra of Random Operators', *Uspekhi Mat. Nauk*, 1973, 28, no. 1, pp. 3-64. [Russian].

17. Prokhorov, Yu. V. 'Convergence of Random Processes and Limit Probability Theorems', *Teoriya Veroyatnostei i ee Primenenie*, 1956, 1, pp. 177-238. [Russian].

18. Sazanov, V. V. 'Remark on Characteristic Functionals', *Teoriya Veroyatnostei i ee Primenenie*, 1958, 3, pp. 201-205. [Russian].

19. Sazonov, V. V., Tutubalin, V. N. 'Probability Distribution on Topological Groups', *Teoriya Veroyatnostei i ee Primenenie*, 1966, 11, pp. 3-55. [Russian].

20. Skorohod, A. V. *An Investigation into the Theory of Random Processes*, Kiev, Izd. Kiev Univers., 1961, 212 pp. [Russian].

21. Skorohod, A. V. 'Multiplicative Matrix Random Processes', Proc. VII *All-Union Conference on the Theory of Probability and Mathematical Statistics*, Tbilisi, 1963, pp. 81-85. [Russian].

22. Skorohod, A. V. 'Construction of Markow Processes with the Aid of Multiplicative Functionals', in *Winter School on Probability Theory*, Kiev, Inst. Matematiki Akad. Nauk UkrSSR, 1964, pp. 191-216. [Russian].

23. Skorohod, A. V. *Integration in Hilbert Space*. Moscow,

'Nauka', 1975. 232 pp. [Russian].

24. Skorohod, A. V. 'Martingales and Stochastic Semigroups', *Teoriya Sluchainykh Protsessov*, 1976, no. 4, pp. 86-94. [Russian].

25. Skorohod, A. V. 'On Spectral Decomposition for Generalized Random Operators', in *Limit Theorems for Random Processes*. Kiev, 1977. [Russian].

26. Skorohod, A. V. 'Infinite Products of Random Matrices', *Teoriya Sluchainykh Protsessov*, 1977, no. 5, pp. 86-92. [Russian].

27. Smolyanov, O. G., Fomin, S. V. 'Measures on Topological Linear Spaces', *Uspekhi Mat. Nauk*, 1976, 31, no. 4, pp. 3-56. [Russian].

28. Partasarathy, K. R. *Probability Measures on Metric Spaces*. New York, Academic Press, 1967. 276 pp.

29. Skorohod, A. V. *Random Operators in Hilbert Space. Lecture Notes in Mathematics* 330. Berlin-Heidelberg-New York, Springer, 1976. pp. 562-591.